Lecture Notes in Computer Science 11996

More information about this series at http://www.springer.com/series/7412

Maria De Marsico · Gabriella Sanniti di Baja ·
Ana Fred (Eds.)

Pattern Recognition Applications and Methods

8th International Conference, ICPRAM 2019
Prague, Czech Republic, February 19–21, 2019
Revised Selected Papers

 Springer

Editors
Maria De Marsico
Department of Computer Science
Sapienza University of Rome
Rome, Italy

Gabriella Sanniti di Baja
ICAR
Consiglio Nazionale delle Ricerche
Naples, Italy

Ana Fred
Instituto de Telecomunicações
Lisbon University
Lisbon, Portugal

Instituto Superior Técnico
Lisbon University
Lisbon, Portugal

ISSN 0302-9743   ISSN 1611-3349   (electronic)
Lecture Notes in Computer Science
ISBN 978-3-030-40013-2   ISBN 978-3-030-40014-9   (eBook)
https://doi.org/10.1007/978-3-030-40014-9

LNCS Sublibrary: SL6 – Image Processing, Computer Vision, Pattern Recognition, and Graphics

This Springer imprint is published by the registered company Springer Nature Switzerland AG
The registered company address is: Gewerbestrasse 11, 6330 Cham, Switzerland

Preface

The present book includes the extended and revised versions of a set of selected papers from the 8th International Conference on Pattern Recognition Applications and Methods (ICPRAM 2019), held in Prague, Czech Republic, during February 10–21, 2019.

Since its first edition, the aim of the ICPRAM series of conferences has been to represent a meeting point between the recent theoretical research results and applications whose design the novel methods underlie. This is reflected in the structure of this book. The first four papers deal with theory and methods and the next three deal with applications. Being ICPRAM, a general-interest conference, the topics in both sections are quite varied and can meet the interest of many readers.

ICPRAM 2019 received 138 paper submissions from 36 countries, of which the best 6% were included in this book. The papers were selected by the event chairs and their selection relies on a number of criteria. These include the classifications and comments provided by the Program Committee members, the session chairs' evaluation of the quality of presentation and the raised interest, and also the program chairs' global view of all papers included in the technical program. The authors of the selected papers were invited to submit a revised and extended version of their papers having at least 30% innovative material, e.g., new experimental results, further comparisons with competing methods, or deeper details on the adopted strategies and choices.

The first paper in the group characterized by theories and methods is "Fourier Spectral Domain Functional Principal Component Analysis of EEG Signals," by Shengkun Xie and Anna T. Lawniczak. The authors propose a novel approach to the analysis of both signals containing a deterministic component, and of random signals, given that the latter first undergo a suitable transformation. The process first transforms the EEG signals in the Fourier spectral domain, and the extracts their key features fusing Functional Principal Component Analysis (FPCA). The experiments in epilepsy diagnosis and epileptic seizure detection demonstrate that the preliminary transformation of EEG signals into their Fourier power spectra significantly enhance the results. In this way, the application of FPCA becomes much more meaningful in signal feature extraction.

In the paper "Online Budgeted Stochastic Coordinate Ascent for Large-Scale Kernelized Dual Support Vector Machine Training," Sahar Qaadan, Abhijeet Pendyala, Merlin Schüler, and Tobias Glasmachers face the problem of the expensive training necessary when using Non-linear Support Vector Machines. A popular strategy to tackle it, especially in the presence large training sets, entails using approximate solvers. An approach achieving good performance requires setting up an a-priori limit on the number of support vectors (the model's budget). The authors revisit recent advances on budgeted training and extend them with several novel components to construct an extremely efficient training algorithm for large-scale data.

The paper "Attributes for Understanding Groups of Binary Data," by Arthur Chambon, Frédéric Lardeux, Frédéric Saubion, and Tristan Boureau deals with relevant attributes for multi-class discrimination of binary data. The set of observations is first divided into groups according to the presence or absence of a set of attributes. Afterward, the authors identify a subset of attributes to describe the groups and present an approach to optimize the choice of the important attributes. The experiments exploit real biological instances to compare the results with competing methods.

In the paper "Annealing by Increasing Resampling," Naoya Higuchi, Yasunobu Imamura, Takeshi Shinohara, Kouichi Hirata, and Tetsuji Kuboyama introduce a stochastic hill-climbing optimization algorithm in which resampling with increasing size relies on the evaluation of the objective function. The algorithm starts as a random walk, while in the ending stages it behaves as a local search. The authors present a unified view of Simulated Annealing (SA) and Annealing by Increased Resampling (AIR), and then generalize both of them to a stochastic hill-climbing for objective functions with stochastic fluctuations, i.e., logit and probit functions, respectively. Since logit can be approximated by probit, AIR can be regarded as an approximation of SA. The authors show the experimental results to compare SA and AIR for sparse pivot selection for dimension reduction and for annealing-based clustering.

The first paper in the second group is "Implications of Z-Normalization in the Matrix Profile," by Dieter De Paepe, Diego Nieves Avendano, and Sofie Van Hoecke. The authors tackle the need of companies to measure their products and services through the increasing available amount of time series data. This results in the equally increasing need of new techniques to extract the requested information. The proposal in the paper starts from adopting the Matrix Profile, which is one state-of-the-art technique for time series and has already been used for various applications including motif/discord discovery, visualizations, and semantic segmentation. After describing it, the authors present a straightforward extension of the distance calculation in the Matrix Profile that benefits from the use of z-normalized Euclidean distance when dealing with time series containing flat and noisy subsequences.

The proposal in the paper "Enforcing the General Planar Motion Model - Bundle Adjustment for Planar Scenes," by Marcus Valtonen Örnhag and Mårten Wadenbäck, tackles the case of planar motion, where a mobile platform equipped with two cameras moves freely on a planar surface. The problem conditions, i.e., the cameras oriented towards the floor but connected to a rigid body in motion, raises new geometrical constraints. The authors propose a bundle adjustment algorithm tailored for the specific problem geometry and also an optimization of the computational demand based on the sparse nature of the problem. The tests of the method on both synthetic and real data provide promising results. On real data, the proposed method shows an improvement compared to generic methods not enforcing the general planar motion model.

Finally, the paper "Deep Multi-Biometric Fusion for Audio-Visual User Re-Identification and Verification," by Mirko Marras, Pedro A. Marín-Reyes, Javier Lorenzo-Navarro, Modesto Castrillon-Santana, and Gianni Fenu illustrates a multi-biometric model training strategy that digests face and voice traits in parallel. The training of a deep network exploits the relation between voice characteristics and facial morphology, so that the two uni-biometric models help each other to recognize people when trained jointly. The authors explore how it can help in improving recognition performance in re-identification and verification scenarios. The experiments on four real-world datasets show a better performance when a uni-biometric model is jointly trained with respect to the case when the same uni-biometric model is trained alone.

ICPRAM would like to become a major point of contact between researchers, engineers, and practitioners in the multi-faceted areas of pattern recognition, both from theoretical and application perspectives. The Organizing Committee especially encourages contributions describing applications of pattern recognition techniques to real-world problems, interdisciplinary research, and experimental and/or theoretical studies yielding new insights that advance pattern recognition methods. This book is a demonstration that this goal can be achieved.

Last but not least, we would like to thank all the authors for their contributions and also to the reviewers who have helped ensure the quality of this publication.

February 2019

Maria De Marsico
Gabriella Sanniti di Baja
Ana Fred

Organization

Conference Chair

Ana Fred — Instituto de Telecomunicações, Instituto Superior Técnico, and Lisbon University, Portugal

Program Co-chairs

Maria De Marsico — Sapienza Università di Roma, Italy
Gabriella Sanniti di Baja — Italian National Research Council (CNR), Italy

Program Committee

Andrea Abate — University of Salerno, Italy
Ashraf AbdelRaouf — Misr International University (MIU), Egypt
Rahib Abiyev — Near East University, Turkey
Gady Agam — Illinois Institute of Technology, USA
Lale Akarun — Bogazici University, Turkey
Mayer Aladjem — Ben-Gurion University of the Negev, Israel
Javad Alirezaie — Ryerson University, Canada
Emili Balaguer-Ballester — Bournemouth University, UK
Stefano Berretti — University of Florence, Italy
Monica Bianchini — University of Siena, Italy
Andrea Bottino — Politecnico di Torino, Italy
Nizar Bouguila — Concordia University, Canada
Francesca Bovolo — Fondazione Bruno Kessler, Italy
Paula Brito — Universidade do Porto, Portugal
Hans Buf — University of Algarve, Portugal
Javier Calpe — Universitat de València, Spain
Francesco Camastra — University of Naples Parthenope, Italy
Mehmet Celenk — Ohio University, USA
Jocelyn Chanussot — Grenoble Institute of Technology, France
Rama Chellappa — University of Maryland, USA
Chi Hau Chen — University of Massachusetts Dartmouth, USA
SongCan Chen — Nanjing University of Aeronautics and Astronautics, China
Chien-Hsing Chou — Tamkang University, Taiwan, China
Francesco Ciompi — Radboud University Medical Center, The Netherlands
Miguel Coimbra — University of Porto, Portugal
Sergio Cruces — Universidad de Sevilla, Spain
Duc-Tien Dang-Nguyen — Dublin City University, Ireland
Justin Dauwels — Nanyang Technological University, Singapore

Luiza de Macedo Mourelle	State University of Rio de Janeiro, Brazil
Maria De Marsico	Sapienza Università di Roma, Italy
Yago Diez	Yamagata University, Japan
Jean-Louis Dillenseger	Université de Rennes 1, France
Ruggero Donida Labati	Università degli Studi di Milano, Italy
Gianfranco Doretto	West Virginia University, USA
Gideon Dror	The Academic College of Tel-Aviv-Yaffo, Israel
Haluk Eren	Firat University, Turkey
Yaokai Feng	Kyushu University, Japan
Gernot Fink	TU Dortmund, Germany
Giorgio Fumera	University of Cagliari, Italy
Vicente Garcia	Autonomous University of Ciudad Juarez, Mexico
Esteban García-Cuesta	Universidad Europea de Madrid, Spain
James Geller	New Jersey Institute of Technology, USA
Angelo Genovese	Università degli Studi di Milano, Italy
Markus Goldstein	Ulm University of Applied Sciences, Germany
Eric Granger	École de Technologie Supérieure, Canada
Sébastien Guérif	University Paris 13 and SPC, France
Robert Harrison	Georgia State University, USA
Pablo Hennings	Huawei Canada Research Center, Canada
Laurent Heutte	Université de Rouen, France
Kouichi Hirata	Kyushu Institute of Technology, Japan
Sean Holden	University of Cambridge, UK
Geoffrey Holmes	University of Waikato, New Zealand
Su-Yun Huang	Academia Sinica, Taiwan, China
Yuji Iwahori	Chubu University, Japan
Sarangapani Jagannathan	Missouri University of Science and Technology, USA
Nursuriati Jamil	Universiti Teknologi MARA, Malaysia
Yunho Kim	Ulsan National Institute of Science and Technology, South Korea
Constantine Kotropoulos	Aristotle University of Thessaloniki, Greece
Sotiris Kotsiantis	University of Patras, Greece
Konstantinos Koutroumbas	National Observatory of Athens, Greece
Kidiyo Kpalma	INSA de Rennes, France
Marek Kretowski	Bialystok University of Technology, Poland
Adam Krzyzak	Concordia University, Canada
Piotr Kulczycki	Polish Academy of Sciences, Poland
Marco La Cascia	Università degli Studi di Palermo, Italy
Shang-Hong Lai	National Tsing Hua University, Taiwan, China
Nikolaos Laskaris	AUTH, Greece
Young-Koo Lee	Kyung Hee University, South Korea
Nicolas Lermé	Université Paris-Sud, France
Aristidis Likas	University of Ioannina, Greece
Shizhu Liu	Apple, USA
Eduardo Lleida	Universidad de Zaragoza, Spain
Luca Lombardi	University of Pavia, Italy

Bob Zhang	University of Macau, Macau, China
Xiao Zhitao	Tianjin Polytechnic University, China
Reyer Zwiggelaar	Aberystwyth University, UK

Additional Reviewers

Antonio Barba	Universidad Europea de Madrid, Spain
Malgorzata Kretowska	Bialystok University of Technology, Poland
Luis Moreira-Matias	Kreditech, Germany

Invited Speakers

Bram van Ginneken	Radboud University Medical Center, The Netherlands
Michal Irani	Weizmann Institute of Science, Israel
Davide Maltoni	University of Bologna, Italy

Contents

Theory and Methods

Fourier Spectral Domain Functional Principal Component Analysis of EEG Signals

Shengkun Xie[1][✉] and Anna T. Lawniczak[2]

[1] Department of Global Management Studies, Ted Rogers School of Management,
Ryerson University, Toronto, ON M5B 2K3, Canada
`shengkun.xie@ryerson.ca`
[2] Mathematics and Statistics,
University of Guelph, Guelph, ON N1G 2W1, Canada
`alawnicz@uoguelph.ca`

Abstract. Functional principal component analysis (FPCA) is a natural tool for investigating the functional pattern of functional data. Clustering functional data using FPCA can be an important area in machine learning for signal processing, in particular, for signals that contain a deterministic component. Also, FPCA can be useful for analysis of random signals if an appropriate transformation is applied. In this work, we propose a novel approach by extracting key features of EEG signals in the Fourier spectral domain using FPCA. By first transforming EEG signals into their Fourier power spectra, the functionality of signals is greatly enhanced. Due to this improvement, the application of FPCA becomes much more meaningful in signal feature extraction. Our study shows a great potential of using spectral domain FPCA as a feature extractor for processing EEG signals in both epilepsy diagnosis and epileptic seizure detection.

Keywords: Fourier power spectrum · Functional principal component analysis EEG · Epilepsy diagnosis

1 Introduction

Feature extraction of high-dimensional data has been an important research area in machine learning [2,7,8,27]. The traditional problem of feature extraction of high-dimensional data focuses on the study of multivariate data with a goal of reducing the dimensionality from the original observation domain to a feature domain. This problem is typically encountered in two machine learning tasks, clustering and classification, depending on whether the data labelling is available or not. Clustering is an unsupervised learning technique that aims at grouping the multivariate data into a selected number of clusters, while classification focuses on predicting the group membership of test data by first learning a mathematical model based on the training data and then using this model to make

© Springer Nature Switzerland AG 2020
M. De Marsico et al. (Eds.): ICPRAM 2019, LNCS 11996, pp. 3–22, 2020.
https://doi.org/10.1007/978-3-030-40014-9_1

a prediction. The performance of clustering or classification is often affected by the dimension of multivariate data. This is why the reduction of data dimension is firstly considered in any given problem. Random signals such as EEG or financial time series are special type of high- dimensional data that often contain no deterministic patterns. Instead, the pattern may appear to be stochastic, which is often time dependent if the random signals are observations over time. Moreover, in real-world problems, most of the signals are non-stationary, which makes the time dependency difficult to measure because of the instability or containing too much uncertainty. Due to the fact of lacking a stable pattern that can be modelled by some mathematical equations, the clustering or classification of random signals directly within the original time domain is challenging. Therefore, feature extraction of random signals is usually the most important step to meet the success of classification or clustering [1, 4, 18, 22, 23], and the development of new methodology for this type of problem becomes highly desirable.

In real-world applications of biomedical signal classification [5, 9, 13, 14, 17], a low-dimensional and linearly or non-linearly separable feature vector is highly desirable for both, the ease of data visualization in medical devices and the possibility of using simple classification methods, such as a linear classifier or the k-nearest neighbour (k-NN) method. To meet this goal, there is a lot of current research focusing on feature extraction of signals in the time domain using sparse representation of signals [28]. The idea of sparse representation is to approximate a given signal using a set of basis functions and to obtain a number of coefficients related to the approximation. The obtained coefficients of basis functions are used as signal features. The further study of signal characteristics is then based on these extracted features. However, a set of time domain basis functions used for approximating a given signal may not be a good choice for another signal, thus the extracted features may appear to be extremely volatile. This may lead to a low performance when using these extracted features in applications. To overcome this, the focus of sparse representation of signals has been moved from the time domain to other single or multiple domains. Among many published research works, the time-frequency domain decomposition is the most popular one within this type of approach. It decomposes the signal in terms of time and frequency domain components. By extending the analysis from a single domain to multiple domains, the separability of signal features is often significantly improved and the classification based on extracted features in both time and frequency domains will outperform a single domain approach [6, 12].

For sparse representation of signals, either in time domain or time-frequency domain, the objective is to achieve an explainable or low-dimensional feature vector, so that classification or clustering of signals becomes easier in feature domain [24]. When these signals contain deterministic patterns and the signal to noise ratio is high, they are typically referred to functional data. The functional data can be in spatial domain, temporal domain or both. Some examples of functional data include, but are not limited to images, temperature data, and growth curves [15, 16, 20]. To study functional data, functional data analysis (FDA) is a natural tool, due to its capability of capturing stable, and observable deterministic patterns contained in the data. In [25], we have used functional data analysis and applied functional summary statistics, functional probes and functional principal components to both epileptic EEG signals without seizure and non-epileptic signals. We have demonstrated that feature extraction through the functional data analysis is able to produce low-dimensional and explainable features so that signals of different types can be discriminated. In this extended work, we will further present the results using the epileptic EEG signals with seizure onset. We will focus on the study of using functional principle component analysis by providing more technical details. In particular, we will provide on discussion of the effect of number of basis functions retained in functional PCA for clustering. Also, we will demonstrate the application of the proposed method for both the epilepsy diagnosis and epileptic seizure detection [10, 11, 21, 26].

This paper is organized as follows. In Sect. 2, we discuss the proposed methods including Fourier power spectra and functional principal component analysis. In Sect. 3, the analysis of publicly available EEG data and summary of main results are presented. Finally, we conclude our findings and provide further remarks in Sect. 4.

2 Methods

2.1 Fourier Power Spectra

To define the Fourier power spectra of signals, we have to first apply the Fourier transformation. For a given signal X_t of length n, sampled at discrete times, the discrete Fourier transform (DFT) is defined as

$$d(\omega_j) = n^{-1/2} \sum_{t=1}^{n} X_t e^{-2\pi i \omega_j t}, \tag{1}$$

for $j = 0, 1, \ldots, n-1$, and the frequency $\omega_j = j/n$. Transforming the signal by discrete Fourier transform allows to obtain a concentration of signal powers using a small set of more dominant frequencies. This means that one is able to focus on a selected ω_j and its transformed values $d(\omega_j)$ only. The signal powers that correspond to the selected frequencies provide us a first layer of feature extraction, and due to the focus on the selected frequencies, the feature sparsity is achieved. However, direct use of Fourier transformed values for the given signal is very inconvenient, because they are complex numbers. A signal periodogram is then created to avoid this difficulty. The periodogram for each frequency ω_j is defined as

$$I(\omega_j) = \mid d(\omega_j) \mid^2 = \frac{1}{n}\sum_{t=1}^{n}\sum_{s=1}^{n}(X_t - \bar{X})(X_s - \bar{X})e^{-2\pi i \omega_j(t-s)}$$

$$= \frac{1}{n}\sum_{h=-(n-1)}^{n-1}\sum_{t=1}^{n-|h|}(X_{t+|h|} - \bar{X})(X_s - \bar{X})e^{-2\pi i \omega_j h}$$

$$= \sum_{h=-(n-1)}^{n-1}\hat{\gamma}(h)e^{-2\pi i \omega_j h}, \qquad (2)$$

where $\hat{\gamma}(h)$ is the auto-covariance function of time lag h. From the statistical point of view, auto-covariance function plays a role of capturing the stochastic pattern of a given signal, which refers to the time dependence structure. Therefore, our approach is not on extracting deterministic pattern in original time domain. This makes our approach to be significantly different from the sparse representation of signals or time frequency decomposition methods. Instead, we aim at first extracting the key signal features, by transforming them to an auto-covariance function, so that the periodogram can be obtained. Notice that the periodogram $I(\omega_j)$ is just the Fourier transform of the auto-covariance function, which captures the quadratic covariation of the lagged signals in the spectral domain. $I(\omega_j)$ is also called a power spectrum of a signal X_t. Because of its definition, the periodogram captures the distribution of covariation of a signal in the spectral domain. The larger is a value of the periodogram, the more dominant is its corresponding frequency. Thus, the dominant values determine the signal power spectra. Often these more dominant frequencies correspond to smaller frequency values, which implies that local patterns are more significant than the global one, after the signal is transformed into the spectral domain. Because of this, in this work, we will only focus on the analysis of the power spectrum in a sub-interval, i.e. we will analyze only the first 200 frequency values for the given signals.

To illustrate the points stated above, we present a set of results of the power spectra for selected different types of EEG signals in Fig. 1. One can clearly see that the power spectra of the first 200 frequency values behave similarly within each set of data, but their patterns look differently over different frequency values. In particular, we observe that the signal powers for high frequency values are still high for both the signals from the set D and the set E, but they are significantly decayed for both the set B and the set C. This observation inspires us to consider the functional representation of signal power spectra, as the functionality of signals within each set is now significantly enhanced when compared to the original time series plots. However, as we can see from Fig. 1, the power spectra for each set of signals are still very noisy and further smoothing may be required, in order to achieve a better performance on clustering of extracted features.

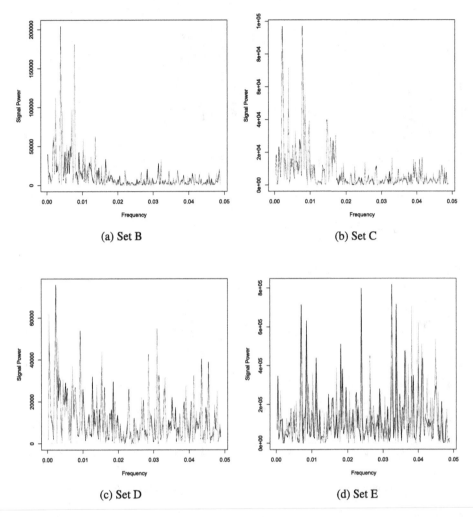

Fig. 1. Sample plots of the power spectra for data sets B (Normal: Eyes Open), C (Non-epileptogenic zone), D (Epileptogenic zone) and E (Seizure onset), respectively. Each plot contains three sample power spectra of the first 200 frequency values. This Figure appears in the proceeding of ICPRAM 2019 [25].

2.2 Functional Representation of Signal Power Spectrum

Our discussion on the signal power spectra in the previous section shows that, they are in fact noisy and further processing is required before we can extract the signal features for clustering. Otherwise, the extracted signal features may be distorted by such noises and a performance of further clustering may not be ideal. Also, a signal is often sampled at a discrete time and further processing is required to make it functional. Equipped with functional data analysis, we can model the power spectrum of a signal with discrete observations by a linear

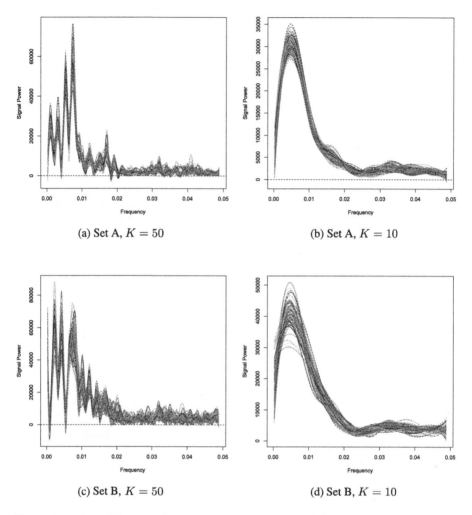

(a) Set A, $K = 50$

(b) Set A, $K = 10$

(c) Set B, $K = 50$

(d) Set B, $K = 10$

Fig. 2. The plots of functional power spectra for data sets A (Normal: Eyes Closed) and B (Normal: Eyes Open), respectively, with different choices of K value. 10 B-splines basis functions are used to smooth sample power spectra.

combination of a set of continuous basis functions. This makes feature extraction of signals to be more mathematically tractable. For a given ith signal, we can expand the power spectrum $I_i(\omega)$ by

$$I_i(\omega) = \sum_{k=1}^{K} \alpha_{ik} \phi_k(\omega), \tag{3}$$

where ω is the frequency value, α_{ik} is the coefficient of the kth basis function and K is the total number of basis functions. Since our objective is not to fully represent a power spectrum using a set of functional basis, K is assumed to be

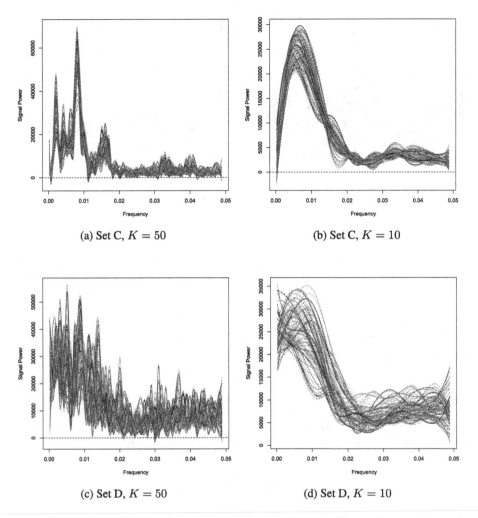

(a) Set C, $K = 50$ (b) Set C, $K = 10$

(c) Set D, $K = 50$ (d) Set D, $K = 10$

Fig. 3. The plots of functional power spectra for data sets C (Non-epileptogenic zone) and D (Epileptogenic zone), respectively, with different choices of K value. 10 B-splines basis functions are used to smooth sample power spectra.

finite and relatively small. That is, we approximate the power spectrum by a linear combination of K basis functions, but for the ease of explanation, we use an equal sign in Eq. (3). However, from the computational perspective, K being finite is always true, as the total number of basis functions needed to approximate the power spectrum is less than the total number of frequency values that we focus on. Notice that, within the discussion of this section, we do not isolate the mean function from the representation of signal. In the later discussion, we will separate the mean function from the signal expansion because of the need of studying functional variation.

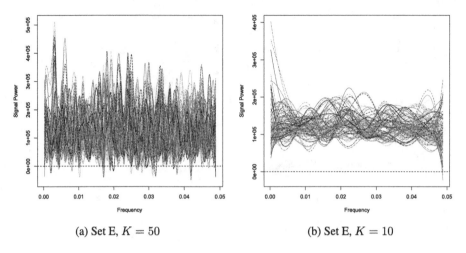

(a) Set E, $K = 50$ (b) Set E, $K = 10$

Fig. 4. The plots of functional power spectra for data sets E (Epileptic Seizure), with different choices of K value. 10 B-splines basis functions are used to smooth sample power spectra.

When considering a sample of N signals, that is $i = 1, 2, \ldots, N$, in the above Eq. (3), the matrix notation of these power spectra becomes

$$\mathbf{I}(\omega) = \mathbf{A}\boldsymbol{\phi}(\omega), \tag{4}$$

where $\mathbf{I}(\omega) = [I_1(\omega), I_2(\omega), \ldots, I_N(\omega)]^\top$ is a column vector of length N and $\boldsymbol{\phi}(\omega) = [\phi_1(\omega), \phi_2(\omega), \ldots, \phi_K(\omega)]^\top$ is a column vector of length K containing the basis functions. $\mathbf{A}$ is the coefficient matrix of the size $N \times K$, i.e.

$$\mathbf{A} = \begin{bmatrix} \alpha_{11} & \alpha_{12} & \alpha_{13} & \cdots & \alpha_{1K} \\ \alpha_{21} & \alpha_{22} & \alpha_{23} & \cdots & \alpha_{2K} \\ \vdots & \vdots & \vdots & \ddots & \vdots \\ \alpha_{N1} & \alpha_{N2} & \alpha_{N3} & \cdots & \alpha_{NK} \end{bmatrix}. \tag{5}$$

Notice that, the set of basis functions $\boldsymbol{\phi}(\omega)$ can be different for various groups of signals, however, given the fact that we will be considering signals that share many commonalities in the spectral domain, it is reasonable to use the same set of basis functions. This allows us to extract signal features within the same feature space. So, in this work, we will hypothesize that signals can be represented by using the same set of basis functions.

2.3   Functional Probes

The descriptive methods such as functional mean, functional variance or functional covariance allow us to see functional central tendency and functional variation patterns. However, they may be functional and high-dimensional as well.

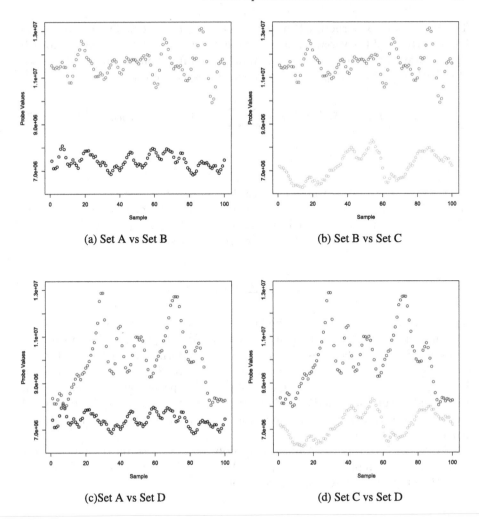

(a) Set A vs Set B

(b) Set B vs Set C

(c)Set A vs Set D

(d) Set C vs Set D

Fig. 5. The plots of extracted probe values using mean function of power spectrum of Set D as the functional probe for data sets A (Normal: Eyes Closed, Black), B (Normal: Eyes Open, Red), C (Non-epileptogenic zone, Green) and D (Epileptogenic zone, Blue), respectively. (Color figure online)

If this is the case, for classification purpose, a dimension of these functional descriptive statistics needs to be further reduced. As a possible dimension reduction method an application of functional probes may be considered. A probe ρ_ξ is a measure that allows us to see specific variation by defining a functional weight $\xi(\omega)$ and it is defined as the inner product of functions $\xi(\omega)$ and $I(\omega)$.

$$\rho_\xi = \int \xi(\omega) I(\omega) d\omega. \tag{6}$$

The $\xi(\omega)$ has to be structured so that we can extract specific features or meaningful patterns of the variation in power spectrum $I(\omega)$. The probe values for ith signal power spectrum using functional mean and functional standard deviation of the jth group power spectrum are defined, respectively, as

$$\rho_{\bar{I}_{ij}} = \int \bar{I}^{(j)}(\omega)I_i(\omega)d\omega = \sum_{k_1=1}^{K}\sum_{k_2=1}^{K} \alpha_{ik_1}\bar{\alpha}_{k_2}^{(j)} \int \phi_{k_1}(\omega)\phi_{k_2}(\omega)d\omega, \qquad (7)$$

$$\rho_{S_{ij}} = \int S_{I_{i(\omega)}}^{(j)} I_i(\omega)d\omega. \qquad (8)$$

The functional probe values capture the similarity between the weight function and the ith power spectrum of a signal. Using functional mean as a probe, when the basis functions are orthogonal, i.e., $\int \phi_{k_1}(\omega)\phi_{k_2}(\omega)d\omega = 0$, for $k_1 \neq k_2$, and $\int \phi_k^2(\omega)d\omega = 1$, for $k = 1, 2, \ldots$, the probe values become

$$\rho_{\bar{I}_{ij}} = \sum_{k=1}^{K} \alpha_{ik}\bar{\alpha}_k^{(j)}, \qquad (9)$$

and they can be interpreted as a similarity measure between two different groups of signals in the spectrum domain, in terms of overall pattern. Unlike the case of using functional mean as a probe, the closed form does not exist for the probe values using functional standard deviation. In this work, we continue to investigate how the extracted signal features (i.e., probe values) behave when functional mean and functional standard deviation are used as the functional probes. However, we focus only on one dimensional quantity. For two dimensional patterns, we refer to [25].

So far, we have discussed functional probe values based on the power spectrum $I(\omega)$. If we replace $I(\omega)$ by $v(\omega_1, \omega_2)$, i.e. by the variance-covariance function, then the functional probe value becomes

$$\int \xi(\omega_2)v(\omega_1, \omega_2)d\omega_2. \qquad (10)$$

This is exactly the left hand side of eigenequation for solving eigenvalues and eigenvectors in functional principal component analysis. More discussion of this type of probe will be provided later.

2.4   Classical Principal Component Analysis

In multivariate statistics, principal component analysis (PCA) of a p-variate random vector $X = (X_1, X_2, \ldots, X_p)$ looks for a set of weight values, denoted by $\xi_j = (\xi_{1j}, \xi_{1j}, \ldots, \xi_{pj})$, so that, at the jth step, the variance of linear combinations of variable X_i is maximized. That is,

$$Var\left(\sum_{i=1}^{p} \xi_{ij} X_i^{(j)}\right) \qquad (11)$$

is maximized or in matrix notation, $Var(X^{(j)}\xi_j^\top)$ is maximized, where

$$X^{(j)} = X^{(j-1)} - \xi_{j-1}^\top X^{(j-1)}, \text{ for } j = 1, 2, \ldots, p, \tag{12}$$

where $X^{(0)}$ is defined as $\mathbf{0}$. This process is repeated for $j = 1, 2, \ldots, p$, and at each step, it is subject to

$$\sum_{i=1}^p \xi_{ij}^2 = 1 \text{ and } \sum_{i=1}^p \xi_{ij}\xi_{il} = 0, \text{ for } j < l, \text{ and } 1 \leq l, j \leq p.$$

An alternative approach of finding a solution of ξ_j is by using a singular value decomposition (SVD) of the data matrix $\mathbf{X}$, which contains N realizations of X. The solution of ξ_j is the jth eigenvector obtained from the SVD of the data matrix $\mathbf{X}$. In this work, the data matrix $\mathbf{X}$ becomes $\mathbf{I}(\omega)$, which is a $N \times p$ data matrix, where p is the total number of frequency values being considered. Technically, the SVD of $\mathbf{I}(\omega)$ is a factorization of the form $\mathbf{U\Sigma V}$, where $\mathbf{U}$ is a $N \times N$ unitary matrix, Σ is a $N \times p$ diagonal matrix consisting of eigenvalues of $\mathbf{I}(\omega)$ and $\mathbf{V}$ is a $p \times p$ unitary matrix. The columns of $\mathbf{U}$ and the columns of $\mathbf{V}$ are called the left eigenvectors and right eigenvectors of $\mathbf{I}(\omega)$, respectively. Also, each column of $\mathbf{V}$ is just the weight vector ξ_j. The feature extraction of data matrix $\mathbf{I}(\omega)$ becomes a problem of computing $\mathbf{I}(\omega)\xi_j^\top$, for $j = 1, 2, \ldots, p$. For example, the first principal component scores set is $\mathbf{I}(\omega)\xi_1^\top$, and the second principal component scores set is $\mathbf{I}(\omega)\xi_2^\top$, etc.

We should also realize that the objective function in PCA can be rewritten as $\xi_j\mathbf{X}^\top\mathbf{X}\xi_j^\top$, assuming that vector $\mathbf{X}$ is centred. In this case, $\mathbf{X}^\top\mathbf{X}$ becomes the variance - covariance matrix. The vector ξ_j is still the eigenvector that is associated with the variance and covariance matrix. All of these can help us understand the relationships among the functional probe, PCA and the functional PCA, which will be discussed later. Notice that, the functional probes discussed above aim at capturing the variation of data associated with the weight function. If we carefully select the functional weight $\xi(\omega)$, so that, the variance of functional probe values in (6) is maximized, subject to the constraint that $\int \xi_l(\omega)\xi_j(\omega)d\omega = 0$ for $l \neq j$, and $\int \xi_j^2(\omega)d\omega = 1$, then it becomes the functional PCA. In this case, the functional probe values are the principal component scores and the weight function becomes functional principal component loadings.

2.5 Functional Principal Component Analysis

So far, we have explained how functional probe and classical principal component analysis work in analyzing the power spectra of random signals. We have also mentioned that when principal component analysis is combined with functional probe, it leads to functional principal component analysis. That is, functional principal component analysis is a typical statistical analysis procedure that aims at maximizing the functional probe values with the use of variance-covariance matrix as an input to functional probe.

Suppose that the power spectrum of a given signal can be expanded using K basis functions and it is given as follows:

$$I_i(\omega) = \mu(\omega) + \sum_{k=1}^{K} \beta_{ik}\phi_k(\omega), \tag{13}$$

where $\mu(\omega)$ is the functional mean of power spectrum. Here, we consider an approximation of $I_i(\omega)$ rather than exact representation of the function using infinite number of basis functions. Of course, we can not rule out the possibility of having a very large K in representing a signal, but in this work, we will consider a sparse representation of the power spectrum. Also, obtaining the functional mean of power spectrum requires an estimate. Without loss of generality, we can simply take the grand mean over all ω, denoted by μ_0, as an estimate, so that the Eq. (13) becomes

$$I_i(\omega) = \mu_0 + \sum_{k=1}^{K} \beta_{ik}\phi_k(\omega), \tag{14}$$

or in a matrix notation

$$\mathbf{I} - \mu = \mathbf{C}\phi, \tag{15}$$

where $\mu = (\mu_0, \mu_0, \ldots, \mu_0)^{\top}$, $\phi = (\phi_1, \phi_2, \ldots, \phi_K)^{\top}$, and the coefficient matrix $\mathbf{C}$ is $N \times K$ that can be written as a matrix:

$$\mathbf{C} = \begin{bmatrix} \beta_{11} & \beta_{12} & \beta_{13} & \cdots & \beta_{1K} \\ \beta_{21} & \beta_{22} & \beta_{23} & \cdots & \beta_{2K} \\ \vdots & \vdots & \vdots & \ddots & \vdots \\ \beta_{N1} & \beta_{N2} & \beta_{N3} & \cdots & \beta_{NK} \end{bmatrix}.$$

Now, we will describe how to obtain the function principal components and their scores. First, let us denote the variance-covariance function of the power spectra $\mathbf{I}$ by $v(\omega_1, \omega_2)$. In a matrix notation this function is defined as

$$v(\omega_1, \omega_2) = \frac{1}{N-1}\phi^{\top}(\omega_1)\mathbf{C}^{\top}\mathbf{C}\phi(\omega_2).$$

To solve for the eigenfunction, first we have to solve the following eigenequation for an appropriate eigenvalue λ

$$\int v(\omega_1, \omega_2)\xi(\omega_2)d\omega_2 = \lambda\xi(\omega_1). \tag{16}$$

Suppose that the eigenfunction $\xi(\omega)$ has an expansion

$$\xi(\omega) = \sum_{k=1}^{K} b_k\phi_k(\omega), \tag{17}$$

or in matrix notation

$$\xi(\omega) = \phi^{\top}(\omega)\mathbf{b}, \tag{18}$$

where $\mathbf{b} = (b_1, b_2, \ldots, b_K)^\top$. This yields

$$\int v(\omega_1, \omega_2)\xi(\omega_2)d\omega_2 = \frac{1}{N-1}\int \phi^\top(\omega_1)\mathbf{C}^\top\mathbf{C}$$
$$\phi(\omega_2)\phi^\top(\omega_2)\mathbf{b}d\omega_2.$$
$$= \frac{1}{N-1}\phi^\top(\omega_1)\mathbf{C}^\top\mathbf{C}\boldsymbol{\Phi}\mathbf{b}, \tag{19}$$

where $\boldsymbol{\Phi} = \int \phi(\omega)\phi^\top(\omega)d\omega$ is the matrix that is given as follows:

$$\boldsymbol{\Phi} = \begin{bmatrix} \int \phi_1\phi_1 d\omega & \int \phi_1\phi_2 d\omega & \cdots & \int \phi_1\phi_K d\omega \\ \int \phi_2\phi_1 d\omega & \int \phi_2\phi_2 d\omega & \cdots & \int \phi_2\phi_K d\omega \\ \vdots & \vdots & \vdots & \vdots \\ \int \phi_K\phi_1 d\omega & \int \phi_K\phi_2 d\omega & \cdots & \int \phi_K\phi_K d\omega \end{bmatrix}.$$

The eigenequation in (16) becomes

$$\frac{1}{N-1}\phi^\top(\omega)\mathbf{C}^\top\mathbf{C}\boldsymbol{\Phi}\mathbf{b} = \lambda\phi^\top(\omega)\mathbf{b}. \tag{20}$$

Since (20) must hold for all ω, this implies that the following matrix equation must hold

$$\frac{1}{N-1}\mathbf{C}^\top\mathbf{C}\boldsymbol{\Phi}\mathbf{b} = \lambda\mathbf{b}. \tag{21}$$

To obtain the required principal components, we define $\mathbf{u} = \boldsymbol{\Phi}^{1/2}\mathbf{b}$ and the Eq. (21) becomes

$$\frac{1}{N-1}\boldsymbol{\Phi}^{1/2}\mathbf{C}^\top\mathbf{C}\boldsymbol{\Phi}^{1/2}\mathbf{u} = \lambda\mathbf{u}. \tag{22}$$

By solving the symmetric eigenvalue problem for $\mathbf{u}$ in (22), and then computing $\mathbf{b} = \boldsymbol{\Phi}^{-1/2}\mathbf{u}$ to get the eigenfunction $\xi(\omega)$, we get that

$$\xi(\omega) = \phi^\top(\omega)\boldsymbol{\Phi}^{-1/2}\mathbf{u}. \tag{23}$$

Solving the eigenvalue problem in (22) will produce K different eigenfunctions and their corresponding eigenvalues which we denote by $(\lambda_j, \xi_j(\omega))$.

If $\phi_k(\omega)$ are orthonormal, then $\boldsymbol{\Phi}$ becomes the $K \times K$ identity matrix and the eigenanalysis of the functional PCA problem in (21) reduces to

$$\frac{1}{N-1}\mathbf{C}^\top\mathbf{C}\mathbf{b} = \lambda\mathbf{b},$$

which is the multivariate PCA that replaces variance-covariance matrix by the coefficient matrix $\mathbf{C}$ obtained from the function approximation of power spectrum. However, one should realize that this is not a standard multivariate PCA. From the discussion above, a multivariate PCA conducts eigenanalysis for a $p \times p$ covariance matrix. With function approximation using K basis functions, the eigenanalysis of functional PCA is applied to $K \times K$ coefficient matrix, which depends on the value of K. In case of sparse approximation, which gives a small value of K, solving an eigenequation is much more efficient from the point of view of computational complexity.

2.6   B-Spline as Functional Basis

In this work, we choose the spline function as functional basis which is called B-spline [3,19]. B-spline has been used very often in numerical analysis to approximate a function or a surface. Assume that there is a sequence of non-decreasing real numbers (i.e., $t_k \leq t_{k+1}$), such that $t_0 \leq t_1 \leq \ldots \leq t_{N-1}$, and N is the length of a signal. We call the set $\{t_k \mid k \in \mathbb{Z}\}$ a knot set and each value t_k is referred to as a knot. Next, we define the augmented knot set $t_{-v+2} = \ldots = t_0 \leq t_1 \leq \ldots \leq t_{N-1} = t_N \ldots = t_{N+v-2}$, where v is the order of the B-spline. We have appended the lower and upper indexes because of the use of recursive formula of the B-spline. Furthermore,we reset the index, so that the new index in the augmented knot set becomes $k = 0, 1, \ldots, N + 2v - 3$. For each augmented knot t_k, $k = 0, 1, \ldots, N + 2v - 3$, we define B-spline recursively as follows:

$$B_{k,0}(x) = \begin{cases} 1, & \text{if } t_k \leq x < t_{k+1} \\ 0, & \text{otherwise} \end{cases}$$

$$B_{k,j}(x) = \gamma_{k,j+1}(x)B_{k,j}(x) + [1 - \gamma_{k+1,j+1}]B_{k+1,j}(x), \tag{24}$$

for $j = 0, 1, \ldots, v - 1$, where

$$\gamma_{k,j+1}(x) = \begin{cases} \frac{x-t_k}{t_{k+j}-t_k}, & \text{if } t_{k+j} \neq t_k \\ 0, & \text{otherwise} \end{cases}.$$

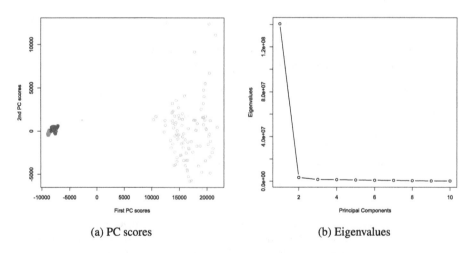

(a) PC scores   (b) Eigenvalues

Fig. 6. The results of principal component scores and eigenvalues of power spectra for C (Non-epileptogenic zone, Green), D (Epileptogenic zone, Blue) and E (Epileptic seizure onset, Light Blue). (Color figure online)

After this recursive process, the maximum number of non-zero basis functions is N, and they are denoted by $B_{1,v-1}, B_{1,v-1}, \ldots, B_{N,v-1}$, and they are called

B-spline basis functions of $v - 1$ degree. If the number of interior points of the knot set, denoted by N^*, is smaller than $N - 2$, then the total number of basis functions is $K = N^* + v$. When $v = 4$, these functions are called cubic B-spline basis functions. In this work, we assume

$$\phi_k(a) = B_{k,3}(a), \text{for } k = 1, 2, \ldots, N^* + 4.$$

So, the maximum number of basis functions used in this work is $K = N^* + 4$.

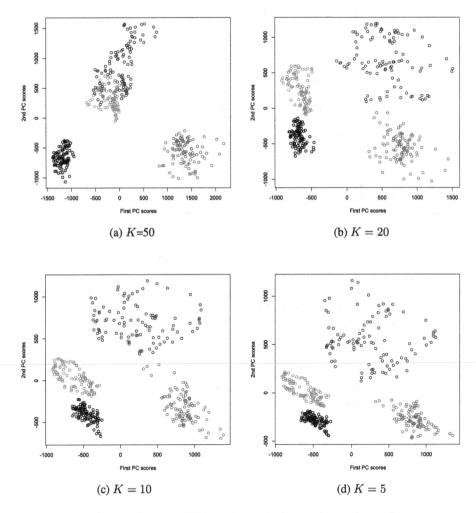

(a) $K = 50$

(b) $K = 20$

(c) $K = 10$

(d) $K = 5$

Fig. 7. The evolution of extracted first two principal component scores of power spectra under different choices of K for data sets A (Normal: Eyes Closed, Black), B (Normal: Eyes Open, Red), C (Non-epileptogenic zone, Green) and D (Epileptogenic zone, Blue). (Color figure online)

2.7 Feature Extraction by Functional Principal Component Analysis

After the jth eigenfunction $\xi_j(\omega)$ is obtained, we can extract the principal component scores, denoted by P_j, for the given power spectrum $\mathbf{I}(\omega)$ as follows

$$P_j = \int \mathbf{I}(\omega)\xi_j(\omega)d\omega, \quad j = 1,\ldots,K. \tag{25}$$

Subsitituting (15) and (23) to the equation above, we get

$$P_j = \int (\mu + \mathbf{C}\phi(\omega))\phi^\top(\omega)\mathbf{\Phi}^{-1/2}\mathbf{u}_j d\omega.$$
$$= \int \mu\phi^\top \mathbf{\Phi}^{-1/2}\mathbf{u}_j + \int \mathbf{C}\phi\phi^\top\mathbf{\Phi}^{-1/2}\mathbf{u}_j$$
$$= \mu\mathbf{\Phi}^{-1/2}\mathbf{u}_j + \mathbf{C}\mathbf{\Phi}^{-1/2}\mathbf{u}_j.$$

Thus, P_1 is the first principal component score vector of the N power spectra of signals, and P_2 is the second principal component score vector, and so on. One of the strengths of using PCA is retaining a small number of principal components, so that, the dimension of feature vector is low. Using the functional PCA on the Fourier power spectra, we have been able to focus on only the first two PCs for clustering the EEG signals.

3 Results

In this study, we use the same data set as in [25], which is from the University of Bonn, Germany (http://epileptologie-bonn.de/cms/front_content.php?idcat=193), but we extend our study by including the set E. We focus on both the epilepsy diagnosis and the epilepsy seizure detection problems. From the results displayed in Figs. 2 and 3, one can see that the numbers of basis functions have a significant effect on the shape of the power spectra. When $K = 10$ is used to smooth out the power spectra, then the smoothed signal power spectra behave similarly for both types of signals, one coming from healthy people (sets A and B) and another one coming from patients for which the signals were collected from a non-epileptogenic zone (set C). However, there are still some differences that we can see among the graphs. This may suggest that further classification is needed based on these power spectra to recognize the differences hidden in the power spectra. Also, one can see that the power spectra of signals collected from patients' epileptogenic zone (set D) are more volatile and look different from the signals of healthy people. However, they share some commonalities with signals from the set C. These differences become more clear when the number of basis functions is $K = 50$. It is understood that when the values of K increase, the smoothed power spectra tend to capture more of the local patterns. This makes a significant difference between different sets of signals. Furthermore, the patterns of power spectra associated with signals from patients with seizure onset (set E) are completely different from other types of signals. From Fig. 4, one can see that there is no clear functionality in the power spectra and with the decrease of K values the overall pattern among the power spectra is still not recognizable. This may suggest the ease of clustering this type of signals from the others. Overall, the graphical display offers some evidence that suitable clustering methods may differentiate these types of signals successfully. In particular, we can aim for the clustering between the sets A and B, this allows us to see the differences caused by artifacts. We can also combine the signals from healthy people, i.e. the sets A and B, and combine the signals from patient without seizure onset, i.e. the sets C and D, in order to see if there is a clustering effect between patients' EEGs and healthy people's EEGs. However, one question still remains. Should the clustering be based on the one that leads to significant overall difference, or the one that offers big difference in local patterns within the power spectra? This question will be answered later in the analysis of the evolution of extracted features.

To further reduce the dimensionality of the power spectrum and its functional mean and its functional standard deviation, the functional probe values are calculated based on the inner product of a selected functional mean and a given signal power spectrum. The results using functional standard deviation as a weight function are also studied. In our study, the best results, in terms of separability of features, are the ones that use the functional mean calculated from the set D. Using the functional mean as a probe, we extract a single-dimensional feature vector from a given signal power spectrum. Figure 5 clearly display the pattern, which shows a great separability of extracted features (i.e., functional probe values), due to the dimension reduction. We observe that the artifacts (i.e., eyes open or eyes closed) associated with the healthy people can be identified when mapping the power spectra of signals on the power spectra of epileptic signals (i.e., set D signals). Using only a single dimension of features, the separation of signals can be achieved for the set A and the set B. Similarly, these single dimension features are highly separable for other cases. This may suggest that the functional probe approach has certain merits in automatic clustering of different types of EEG signals.

To demonstrate the application to epileptic seizure detection, we use functional PCA to extract the principal components of the power spectra of signals from the sets C, D and E. From the results one can see if there exist some dominant signal components. Additionally, the corresponding scores can be used for clustering. The obtained results are displayed in Fig. 6. From Fig. 6(a), one can see that the extracted features for non-seizure signals (sets C and D) and seizure signals (set E) form into clusters and they are linearly separable in first PC. Figure 6(b) display that, the first eigenvalue is very dominant, which explains why the extracted features are highly separable in the first PC. This implies that the extracted first PC scores can be successfully classified using a simple classification method, such as k-NN. The second PC scores are not helpful in contributing to the cluster effect as they are completely overlapped. We further investigate the effect of the number of basis functions (i.e., K) on the separability of extracted signal features (i.e., principal component scores of power spectra). The proposed method is applied to signals from sets A, B, C and D only by varying the value of K. The obtained results are displayed in Fig. 7. We observe that the proposed method is highly successful in separating the artifacts (i.e., open/closed eyes) as the results did not depend on how the number of basis functions was selected. The feature separability increases with the decrease of K, i.e. the number of basis functions. This may suggest that the sparsity in approximation of the signal power spectra plays an important role in the success of applying functional principal component analysis. When $K = 50$, the extracted features for epileptic signals overlap significantly. This overlapping changes when K decreases, and features start to be fully separable when K is relatively small, for example, around 20. When $K = 5$, the obtained results are considered to be optimal in the sense of feature separability. However, the overall separability between healthy and epileptic signals is not affected by the number of basis functions. This may suggest that applying the proposed methods to the epilepsy diagnosis problem can be a successful tool, which allows separating the healthy and patient signal.

4   Concluding Remarks

Clustering and classification of high-dimensional data are important aspects in both pattern recognition and artificial intelligence. To be successful in dealing with clustering and classification, dimension reduction of high-dimensional data plays an important role. In this work, we focus on the study of using feature extraction as a dimension reduction approach. We use EEG signals to illustrate our proposed methodologies. We first transform EEG signals to the spectral domain to obtain their power spectra. Next, we apply functional principal component analysis, which is considered to be a special case of functional probe, to further investigate the characteristics of the signals and to extract their features for clustering. We have demonstrated that functional principal component analysis in spectral domain is useful for better understanding of different types of EEG signals. Furthermore, the extracted features can be used for signal classification. We have also investigated the effect of sparsity on the performance of separating signal features. We have observed that the separability of extracted features is significantly improved, when the number of signal components used for approximating the power spectra decreases. This may imply that the sparse approximation for signal approximation in spectral domain is a necessary step for better performance of signal clustering. From an application perspective, the obtained results demonstrate that the proposed method may be useful for both epilepsy diagnosis and epileptic seizure detection. Future work will focus on the study of wavelet spectral domain functional PCA and its application to clustering general random signals such as financial time series.

References

1. Alickovic, E., Kevric, J., Subasi, A.: Performance evaluation of empirical mode decomposition, discrete wavelet transform, and wavelet packed decomposition for automated epileptic seizure detection and prediction. Biomed. Signal Process. Control **39**, 94–102 (2018)
2. Bouveyron, C., Girard, S., Schmid, C.: High-dimensional data clustering. Comput. Stat. Data Anal. **52**(1), 502–519 (2007)
3. De Boor, C.: A Practical Guide to Splines, vol. 27. Springer, New York (1978)
4. Fergus, P., Hussain, A., Hignett, D., Al-Jumeily, D., Abdel-Aziz, K., Hamdan, H.: A machine learning system for automated whole-brain seizure detection. Appl. Comput. Inf. **12**(1), 70–89 (2016)
5. Gandhi, T., Panigrahi, B.K., Anand, S.: A comparative study of wavelet families for EEG signal classification. Neurocomputing **74**(17), 3051–3057 (2011)
6. Garcia, G.N., Ebrahimi, T., Vesin, J.M.: Support vector EEG classification in the Fourier and time-frequency correlation domains. In: First International IEEE EMBS Conference on Neural Engineering, 2003. Conference Proceedings, pp. 591–594. IEEE (2003)
7. Jimenez, L.O., Landgrebe, D.A.: Supervised classification in high-dimensional space: geometrical, statistical, and asymptotical properties of multivariate data. IEEE Trans. Syst. Man Cybern. Part C (Appl. Rev.) **28**(1), 39–54 (1998)
8. Kriegel, H.P., Kröger, P., Zimek, A.: Clustering high-dimensional data: a survey on subspace clustering, pattern-based clustering, and correlation clustering. ACM Trans. Knowl. Discov. Data (TKDD) **3**(1), 1 (2009)
9. Li, D., Pedrycz, W., Pizzi, N.J.: Fuzzy wavelet packet based feature extraction method and its application to biomedical signal classification. IEEE Trans. Biomed. Eng. **52**(6), 1132–1139 (2005)
10. Liang, S.F., Wang, H.C., Chang, W.L.: Combination of EEG complexity and spectral analysis for epilepsy diagnosis and seizure detection. EURASIP J. Adv. Signal Process. **2010**, 62 (2010)
11. Lima, C.A., Coelho, A.L., Chagas, S.: Automatic EEG signal classification for epilepsy diagnosis with relevance vector machines. Expert Syst. Appl. **36**(6), 10054–10059 (2009)
12. Nyan, M., Tay, F., Seah, K., Sitoh, Y.: Classification of gait patterns in the time-frequency domain. J. Biomech. **39**(14), 2647–2656 (2006)
13. Phinyomark, A., Phukpattaranont, P., Limsakul, C.: Feature reduction and selection for EMG signal classification. Expert Syst. Appl. **39**(8), 7420–7431 (2012)
14. Qazi, K.I., Lam, H., Xiao, B., Ouyang, G., Yin, X.: Classification of epilepsy using computational intelligence techniques. CAAI Trans. Intell. Technol. **1**(2), 137–149 (2016)
15. Ramsay, J.: Functional Data Analysis. Encyclopedia of Statistics in Behavioral Science (2005)
16. Ramsay, J.O., Silverman, B.W.: Applied Functional Data Analysis: Methods and Case Studies. Springer, New York (2007)
17. Subasi, A., Gursoy, M.I.: EEG signal classification using PCA, ICA, LDA and support vector machines. Expert Syst. Appl. **37**(12), 8659–8666 (2010)
18. Truong, N.D., Kuhlmann, L., Bonyadi, M.R., Yang, J., Faulks, A., Kavehei, O.: Supervised learning in automatic channel selection for epileptic seizure detection. Expert Syst. Appl. **86**, 199–207 (2017)

19. Unser, M., Aldroubi, A., Eden, M., et al.: B-spline signal processing: part I theory. IEEE Trans. Signal Process. **41**(2), 821–833 (1993)
20. Wang, J.L., Chiou, J.M., Müller, H.G.: Functional data analysis. Ann. Rev. Stat. Appl. **3**, 257–295 (2016)
21. Xie, S., Krishnan, S.: Signal decomposition by multi-scale PCA and its applications to long-term EEG signal classification. In: The 2011 IEEE/ICME International Conference on Complex Medical Engineering, pp. 532–537. IEEE (2011)
22. Xie, S., Krishnan, S.: Wavelet-based sparse functional linear model with applications to EEGs seizure detection and epilepsy diagnosis. Med. Biol. Eng. Comput. **51**(1–2), 49–60 (2013)
23. Xie, S., Krishnan, S.: Dynamic principal component analysis with nonoverlapping moving window and its applications to epileptic EEG classification. Sci. World J. **2014**, 1–10 (2014)
24. Xie, S., Krishnan, S.: Model based sparse feature extraction for biomedical signal classification. Int. J. Stat. Med. Res. **6**(1), 10–21 (2017)
25. Xie, S., Lawniczak, A.: Feature extraction of eeg in spectral domain via functional data analysis. In: International Conference on Pattern Recognition Applications and Methods, pp. 118–127 (2019)
26. Xie, S., Lawniczak, A.T., Song, Y., Liò, P.: Feature extraction via dynamic PCA for epilepsy diagnosis and epileptic seizure detection. In: 2010 IEEE International Workshop on Machine Learning for Signal Processing, pp. 337–342. IEEE (2010)
27. Yu, L., Liu, H.: Feature selection for high-dimensional data: a fast correlation-based filter solution. In: Proceedings of the 20th International Conference on Machine Learning (ICML-03), pp. 856–863 (2003)
28. Zhang, Z., Xu, Y., Yang, J., Li, X., Zhang, D.: A survey of sparse representation: algorithms and applications. IEEE Access **3**, 490–530 (2015)

Online Budgeted Stochastic Coordinate Ascent for Large-Scale Kernelized Dual Support Vector Machine Training

Sahar Qaadan$^{(\boxtimes)}$, Abhijeet Pendyala, Merlin Schüler, and Tobias Glasmachers

Institute for Neural Computation, Ruhr University Bochum, 44801 Bochum, Germany
sahar.qaadan@ini.rub.de

Abstract. Non-linear Support Vector Machines yield highly accurate predictions for many types of data. Their main drawback is expensive training. For large training sets, approximate solvers are a must. An a-priori limit on the number of support vectors (the model's budget), which is maintained by successive merging of support vectors, has proven to be a well suited approach. In this paper we revisit recant advances on budgeted training and enrich them with several novel components to construct an extremely efficient training algorithm for large-scale data. Starting from a recently introduced dual algorithm, we improve the search for merge partners, incorporate a fast merging scheme, adaptive coordinate frequency selection, and an adaptive stopping criterion. We provide an efficient implementation and demonstrate its superiority over several existing solvers.

Keywords: Support vector machines · Budget maintenance adaptive coordinate descent · Kernel methods

1 Introduction

The Support Vector Machine (SVM) introduced by [7] is popular machine learning method, in particular for binary classification, which is an important problem class in machine learning. Given a set of training sample vectors x and corresponding labels $y \in Y = \{-1, +1\}$ the task is to estimate the label y' of a previously unseen vector x'. Due to its large margin property is yields excellent results.

Linear support vector machines are scalable to extremely large data sets. For this special case, fast online solvers are available [10,31]. However, many problems are not linearly separable. These problems require kernel-based SVMs [1,12,14,22,37]. They are supported by strong learning theoretical guarantees. Being kernel methods, SVM employs a linear algorithm in an implicitly defined kernel-induced feature space. Unlike their linear variant they suffer from various drawbacks in terms of computational and memory efficiency. Their model is represented as a sum of functions over the set of support vectors, which has

© Springer Nature Switzerland AG 2020
M. De Marsico et al. (Eds.): ICPRAM 2019, LNCS 11996, pp. 23–47, 2020.
https://doi.org/10.1007/978-3-030-40014-9_2

been theoretically [32] and experimentally shown to grow linearly with the size of the training set.

The optimization problem of finding the large margin classifier can be cast as a quadratic program (QP), i.e., a quadratic cost function subject to linear constraints. Off-the-shelf solvers can be applied only to small sized data sets due to its high computational and memory costs. The practical application of SVMs began with the introduction of decomposition methods such as sequential minimal optimization (SMO) [5,6,26,27] and SVM^{light} [15], which apply subspace ascent to the dual representation of the problem. These methods could handle medium sized data sets, large enough at the time, but the runtimes grows quadratic to cubic with the size of the training data, limiting their applicability to larger data sets [2]. The present paper is concerned with one class of approximate training methods addressing this scalability problem, namely with so-called budget methods.

SVMs can be trained by solving their primal or dual optimization problem. While the primal is generally solved with stochastic gradient descent (SGD) [31] and variants thereof [21], solving the dual with decomposition methods is usually more efficient [2]. When training with universal kernels without offset this amounts to coordinate ascent [28,33]. It turns out that the iteration complexity of primal stochastic gradient descent and dual coordinate ascent is very similar: both are governed by the cost of evaluating the model on a training point. This cost is proportional to the number of support vectors, which grows at a linear rate with the data set size [32]. This limits the applicability of kernel methods to large-scale data. Several online algorithms fix the number of SVs to pre-specified value $B \ll n$, the budget, and update the model in a greedy manner. This way they limit the iteration complexity. The budget approach can achieve significant reductions in training time (and in the final number of support vectors) with only minimal sacrifices in predictive accuracy.

There exists a considerable number of budget-based algorithms for the training of kernel classifiers, like the Fixed Budget Perceptron [8], the Random Perceptron [4], the Stoptron [25], the Forgetron [9], Tighter Perceptron [36] and Tightest Perceptron [34], and the Projectron [25]. They implement different budget maintenance strategies. A systematic study of budget maintenance strategies for (primal) SVM training is found in [35], where removal of the support vector (SV) with smallest norm, projection, and merging of SVs are analyzed and compared.

Merging is the state-of-the-art method of choice. It merges two SVs into a new SV that is not necessarily part of the training data. Merging of two support vectors has proven to be a good compromise between the induced error and the resulting computational effort. The time complexity for selecting merge partners is usually $\mathcal{O}(B)$ by using a first-choice heuristic. The application of merging-based budget maintenance to SGD-based primal SVM training has become known as budgeted stochastic gradient descent (BSGD) [35].

This paper is an extended version of our prior work [28]. Against the background detailed above, our contributions are as follows:

- We introduce a dual training algorithm with budget constraint [28]. The approach combines the fast convergence behavior of dual decomposition algorithms with fast iterations of the budget method. We call the new solver budgeted stochastic coordinate ascent (BSCA).
- We provide a theoretical analysis of the convergence behavior of BSCA.
- We empirically demonstrate its superiority over primal BSGD.

The dual solver presented in [28] revers to rather simplistic techniques compared to the advanced variable selection and stopping rules employed by exact (non-budgeted) dual solvers [2]. That was necessary because a fast budget solver cannot keep track of the dual gradient. By addressing these issues we go beyond [28] as follows:

- We incorporate the adaptive coordinate frequency selection scheme [11] into the new dual solver to achieve further speed-ups.
- We further reduce the effort spent on merging and in particular on the selection of merging candidates by introducing a novel sampling scheme, and by incorporating the fast lookup method [13].
- Stopping criteria based on Karush-Kuhn Tucker (KKT) violations do not work for the new solver due to the budget constraint. Therefore we introduce a new stopping heuristic.
- We present additional experiments evaluating these contributions.

The paper is organized as follows. In the next section we briefly introduce SVMs and the budgeting technique. We proceed by presenting our dual solver with budget constraint. In Sect. 5 the adaptive coordinate selection technique is added to the solver, followed by further speed-up techniques. The new stopping heuristic is proposed in Sect. 7. An extensive experimental evaluation is undertaken in Sect. 8. We close with our conclusions.

2   Support Vector Machine Training

A Support Vector Machine is a machine learning algorithm for data classification [7]. Given a set of instance-label pairs $(x_1, y_1), \ldots, (x_n, y_n) \in X \times Y$ and a kernel function $k : X \times X \to \mathbb{R}$ over the input space, obtaining the optimal SVM decision function[1] $f(x) \mapsto \langle w, \phi(x) \rangle$ requires solving the following unconstrained optimization problem

$$\min_{w} \quad P(w) = \frac{1}{2}\|w\|^2 + \frac{C}{n}\sum_{i=1}^{n} L\big(y_i, f(x_i)\big), \tag{1}$$

where $\lambda > 0$ is a regularization parameter, and L is a loss function (usually convex in w, turning problem (1) into a convex problem). Kernel methods map training points $x_i \in X$ into a high-dimensional space $\mathcal{H}$ through a non-linear function $\phi(x)$ fulfilling $\langle \phi(x), \phi(x') \rangle = k(x, x')$. Here $k : X \times X \to \mathbb{R}$ is a (Mercer)

---

[1] In this work we consider SVMs without bias, see [33].

kernel function. Replacing x_i with $\phi(x_i)$ in problem (1) and solving the problem for $w \in \mathcal{H}$ we obtain a non-linear or kernelized SVM. The representer theorem allows to restrict the solution to the form $w = \sum_{i=1}^{n} \alpha_i y_i \phi(x_i)$ with coefficient vector $\alpha \in \mathbb{R}^n$, yielding $f(x) = \sum_{i=1}^{n} \alpha_i y_i k(x, x_i)$. Training points x_i with non-zero coefficients $\alpha_i \neq 0$ are called support vectors.

Problem (1) is often referred to as the primal form of SVM. One may instead solve its dual problem [2]:

$$\max_{\alpha \in [0,C]^n} \quad D(\alpha) = \mathbb{1}^T \alpha - \frac{1}{2} \alpha^T Q \alpha, \tag{2}$$

which is a box-constrained quadratic program (QP), with $\mathbb{1} = (1, \ldots, 1)^T$ and $C = \frac{1}{\lambda n}$. The matrix Q consists of the entries $Q_{ij} = y_i y_j k(x_i, x_j)$.

2.1   Non-linear SVM Solvers

Dual decomposition solvers like LIBSVM [2,5] are especially designed for obtaining a high-precision non-linear (kernelized) SVM solution. They work by decomposing the dual problem into a sequence of smaller problems of size $q = \mathcal{O}(1)$, in each iteration, only q columns of the Hessian matrix Q are required. They can be calculated and stored in the computer memory when needed. Unlike other optimization methods which usually require the access of the whole Q.

For problem (2) this can amount to coordinate ascent (CA). Keeping track of the dual gradient $\nabla_\alpha D(\alpha) = \mathbb{1} - Q\alpha$ allows for the application of elaborate heuristics for deciding what coordinate to optimize next, based on the violation of the Karush-Kuhn-Tucker conditions or even taking second order information into account. Provided that coordinate i is to be optimized in the current iteration, the sub-problem restricted to α_i is a one-dimensional QP, which is solved optimally by the truncated Newton step

$$\alpha_i \leftarrow \left[\alpha_i + \frac{1 - Q_i \alpha}{Q_{ii}} \right]_0^C, \tag{3}$$

where Q_i is the i-th row of Q and $[x]_0^C = \max\{0, \min\{C, x\}\}$ denotes truncation to the box constraints. The method enjoys locally linear convergence [18], polynomial worst-case complexity [19], and fast convergence in practice.

In principle the primal problem (1) can be solved directly, e.g., with SGD, which is at the core of the kernelized Pegasos algorithm [31]. Replacing the average loss (empirical risk) in Eq. (1) with the loss $L(y_i, f(x_i))$ on a single training point selected uniformly at random provides an unbiased estimate. Following its (stochastic) sub-gradient with learning rate $1/(\lambda t) = (nC)/t$ in iteration t yields the update

$$\alpha \leftarrow \alpha - \frac{\alpha}{t} + \mathbb{1}_{\{y_i f(x_i) < 1\}} \frac{nC}{t} e_i, \tag{4}$$

where e_i is the i-th unit vector and $\mathbb{1}_{\{E\}}$ is the indicator function of the event E. Despite fast initial progress, the procedure can take a long time to produce accurate results, since SGD suffers from the non-smooth hinge loss, resulting in slow convergence.

In both algorithms, the iteration complexity is governed by the computation of $f(x)$ (or equivalently, by the update of the dual gradient), which is linear in the number of non-zero coefficients α_i. This is a limiting factor when working with large-scale data, since the number of support vectors is usually linear in the data set size n [32].

2.2 Budgeted Stochastic Gradient Descent

Budgeted Stochastic Gradient Descent (BSGD) breaks the unlimited growth in model size and hence in update time for large data sets by bounding the number of support vectors during training. The upper bound $B \ll n$ is the budget size. If B is independent of n then the iteration complexity of BSGD becomes *independent* of the data set size.

Each SGD step can add at most one new support vector. This happens exactly if (x_i, y_i) does not meet the target margin of one, and α_i changes from zero to a non-zero value. The resulting model is of the form

$$\tilde{w} = \sum_{(\beta, \tilde{x}) \in M} \beta \cdot \phi(\tilde{x}) \qquad \text{s.t. } |M| \le B, \tag{5}$$

where the at most B support vectors $\tilde{x}_j$ are not necessarily training points, and $\tilde{w}$ shall approximate w reasonably well under the budget constraint.

After $B + 1$ such steps, the model holds more than B points. Hence, the budget condition is violated and a dedicated budget maintenance algorithm is triggered to reduce the number of support vectors to at most B. The goal of budget maintenance is to fulfill the budget constraint with the smallest possible change of the model, measured by $\|\Delta\|^2 = \|w' - w\|^2$, where w is the weight vector before and w' is the weight vector after budget maintenance. The model change $\Delta = w' - w$ is called weight degradation.

Budget maintenance strategies were investigated in detail in [35]. It turned out that *merging* of two support vectors into a single new point is superior to alternatives like removal of a point and projection of the solution onto the remaining support vectors. Merging was first proposed in [24] as an efficient way to reduce the complexity of an already trained SVM. With merging, the complexity of budget maintenance is governed by the search for suitable merge partners, which is $\mathcal{O}(B^2)$ for all pairs, while it is common to apply the $\mathcal{O}(B)$ heuristic resulting from fixing the point with smallest coefficient α_i as a first partner. This way, the weight degradation of merging is guaranteed to be upper bounded by weight degradation of removing a basis point.

When merging two support vectors x_i and x_j, we aim to approximate $\alpha_i \cdot \phi(x_i) + \alpha_j \cdot \phi(x_j)$ with a new term $\alpha_z \cdot \phi(z)$ involving only a single point z. Since the kernel-induced feature map is usually not surjective, the pre-image of $\alpha_i \phi(x_i) + \alpha_j \phi(x_j)$ under ϕ is empty [3,29] and no exact match z exists. Therefore the weight degradation $\Delta = \alpha_i \phi(x_i) + \alpha_j \phi(x_j) - \alpha_z \phi(z)$ is non-zero.

For the Gaussian kernel $k(x_i, x_j) = \exp(-\gamma \|x_i - x_j\|^2)$ it can be solved with golden section search for z on the line through x_i and x_j, while β_z is obtained analytically. For details of the merging procedure we refer to [35].

3 Budgeted Stochastic Coordinate Ascent

In this section we present our novel approximate SVM training algorithm, see also [28]. At its core it is a dual decomposition algorithm, modified to respect a budget constraint. It is designed so that the iteration complexity is limited to $\mathcal{O}(B)$ operations, and is hence independent of the data set size n. The budgeted dual coordinate ascent solver combines components from decomposition methods [26], dual linear SVM solvers [10], and BSGD [35] into a new algorithm. We call it *Budgeted Stochastic Coordinate Ascent (BSCA)*. Like BSGD, the BSCA algorithm features an a-priori limited iteration complexity following the budget approach, but combined with fast convergence of dual decomposition solvers. Both aspects accelerate the training process, and hence allow to scale SVM training to larger problems than with exact decomposition algorithms, and also with BSGD.

Introducing a budget into a standard decomposition algorithm as implemented in LIBSVM [5] turns out to be non-trivial. Incorporating a budget is rather straightforward on the primal problem (1). The optimization problem is unconstrained, allowing BSGD to replace w represented by α transparently with $\tilde{w}$ represented by flexible basis points $\tilde{x}_j$ and coefficients β_j in Eq. (5). This is not possible for the dual problem (2) with constraints formulated directly in terms of α.

The same difficulty is solved by [10] for the linear SVM training problem by keeping track of w and α in parallel. We follow the same approach, however, the correspondence between w represented by α and $\tilde{w}$ represented by β_j and $\tilde{x}_j$ is

---

Algorithm 1. Budgeted Stochastic Coordinate Ascent (BSCA) Algorithm.

1: **Input:** training data $(x_1, y_1), \ldots, (x_n, y_n)$, $k : X \times X \to \mathbb{R}$, $C > 0$, $B \in \mathbb{N}$
2: $\alpha \leftarrow 0$, $M \leftarrow \emptyset$
3: **while** not happy **do**
4:      select index $i \in \{1, \ldots, n\}$ uniformly at random
5:      $\tilde{f}(x_i) = \sum_{(\beta, \tilde{x}) \in M} \beta k(x_i, \tilde{x})$
6:      $\delta = \left[\alpha_i + \left(1 - y_i \tilde{f}(x_i)\right)/Q_{ii} \right]_0^C - \alpha_i$
7:      **if** $\delta \neq 0$ **then**
8:           $\alpha_i \leftarrow \alpha_i + \delta$
9:           $M \leftarrow M \cup \{(\delta, x_i)\}$
10:          **if** $|M| > B$ **then**
11:               trigger budget maintenance, i.e., merge two support vectors
12:          **end if**
13:      **end if**
14: **end while** return M

---

only approximate. This is unavoidable by the very nature of the budget method being an approximation scheme. Luckily, this does not impose major additional complications.

The pseudo-code of the Budgeted Stochastic Coordinate Ascent (BSCA) approach is provided in Algorithm 1. It represents the approximate model $\tilde{w}$ as a set M containing tuples $(\beta, \tilde{x})$. The budget constraint becomes $|M| \leq B$. Decisively, in line 5 the approximate model $\tilde{w}$ is used to compute $\tilde{f}(x_i) = \langle \tilde{w}, x_i \rangle$, so the complexity of this step is $\mathcal{O}(B)$. This is in contrast to the computation of $f(x_i) = \langle w, x_i \rangle$, with effort linear in n.

At the iteration cost of $\mathcal{O}(B)$ it is not possible to keep track of the dual gradient $\nabla D(\alpha) = \mathbb{1} - Q\alpha$ because it consists of n entries that would need updating with a dense matrix row Q_i. This implies the need for a number of changes compared to exact dual decomposition methods:

- In line with [10], BSCA resorts to uniform variable selection in an SCA scheme.
- The role of the coefficients α is reduced to keeping track of the constraints. We extend this scheme in Sect. 5 below.
- The usual stopping criterion based on the largest violation of the Karush-Kuhn-Tucker (KKT) conditions does not work any longer. We develop a replacement in Sect. 7.

For budget maintenance, the same options are available as in BSGD. It is implemented as merging of two support vectors, reducing a model from size $|M| = B + 1$ back to size $|M| = B$. It is clear that also the complexity of the budget maintenance procedure should be bounded by $\mathcal{O}(B)$ operations. Furthermore, for the overall algorithm to work properly, it is important to maintain the approximate relation $\tilde{w} \approx w$. For reasonable settings of the budget B this is achieved by non-trivial budget maintenance procedures like merging and projection [35].

4   Analysis of BSCA

BSCA is an approximate dual training scheme. Therefore two questions of major interest are

1. how quickly BSCA approaches the optimal solution w^*, and
2. how close does it get?

To simplify matters we make the technical assumption that the matrix Q is strictly positive definite. This ensures that the optimal coefficient vector α^* corresponding to w^* is unique.[2] For a given weight vector $w = \sum_{i=1}^{n} \alpha_i y_i \phi(x_i)$, we write $\alpha(w)$ when referring to the corresponding coefficients, which are also unique. Let $w^{(t)}$ and $\alpha^{(t)} = \alpha(w^{(t)})$, $t \in \mathbb{N}$, denote the sequence of solutions

---

[2] For Gaussian kernels this amounts to the rather weak assumption that training points are unique.

generated by an iterative algorithm, using the training point $(x_{i^{(t)}}, y_{i^{(t)}})$ for its update in iteration t. The indices $i^{(t)} \in \{1, \ldots, n\}$ are drawn i.i.d. from the uniform distribution.

4.1   Optimization Progress of BSCA

The following Lemma computes the single-iteration progress.

Lemma 1. *The change* $D(\alpha^{(t)}) - D(\alpha^{(t-1)})$ *of the dual objective function in iteration t operating on the coordinate index $i = i^{(t)} \in \{1, \ldots, n\}$ equals*

$$J\left(\alpha^{(t-1)}, i, \alpha_i^{(t)} - \alpha_i^{(t-1)}\right)$$
$$:= \frac{Q_{ii}}{2}\left(\left[\frac{1 - Q_i\alpha^{(t-1)}}{Q_{ii}}\right]^2 - \left[\left(\alpha_i^{(t)} - \alpha_i^{(t-1)}\right) - \frac{1 - Q_i\alpha^{(t-1)}}{Q_{ii}}\right]^2\right).$$

Proof. Consider the function $s(\delta) = D(\alpha^{(t-1)} + \delta e_i)$. It is quadratic with second derivative $-Q_{ii} < 0$ and with its maximum at $\delta^* = (1 - Q_i\alpha^{(t-1)})/Q_{ii}$. Represented by its second order Taylor series around δ^* it reads $s(\delta) = s(\delta^*) - \frac{Q_{ii}}{2}(\delta - \delta^*)^2$. This immediately yields the result.

The lemma is in line with the optimality of the update (3). Based thereon we define the relative approximation error

$$E(w, \tilde{w}) := 1 - \max_{i \in \{1, \ldots, n\}} \left\{ \frac{J\left(\alpha(w), i, \left[\alpha_i(w) + \frac{1 - y_i\langle \tilde{w}, \phi(x_i)\rangle}{Q_{ii}}\right]_0^C - \alpha_i(w)\right)}{J\left(\alpha(w), i, \left[\alpha_i(w) + \frac{1 - y_i\langle w, \phi(x_i)\rangle}{Q_{ii}}\right]_0^C - \alpha_i(w)\right)} \right\}. \tag{6}$$

Note that the margin calculation in the numerator is based on $\tilde{w}$, while it is based on w in the denominator. Hence $E(w, \tilde{w})$ captures the effect of using $\tilde{w}$ instead of w in BSCA. It can be (informally) interpreted as a dual quantity related to the weight degradation error $\|\tilde{w} - w\|^2$. Lemma 1 implies that the relative approximation error is non-negative, because the optimal step is in the denominator, which upper bounds the fraction by one. It is continuous (and in fact piecewise linear) in $\tilde{w}$, for fixed w. Finally, it fulfills $\tilde{w} = w \Rightarrow E(w, \tilde{w}) = 0$. The following theorem bounds the suboptimality of BSCA. It formalizes the intuition that the relative approximation error poses a principled limit on the achievable solution precision.

Theorem 1. *The sequence $\alpha^{(t)}$ produced by BSCA fulfills*

$$D(\alpha^*) - \mathbb{E}\left[D(\alpha^{(t)})\right] \le \left(D(\alpha^*) + \frac{nC^2}{2}\right) \cdot \prod_{\tau=1}^{t}\left(1 - \frac{2\kappa\left(1 - E(w^{(\tau-1)}, \tilde{w}^{(\tau-1)})\right)}{(1 + \kappa)n}\right),$$

where κ is the smallest eigenvalue of Q.

Proof. Theorem 5 by [23] applied to the non-budgeted setting ensures linear convergence

$$\mathbb{E}[D(\alpha^*) - D(\alpha^{(t)})] \leq \left(D(\alpha^*) + \frac{nC^2}{2} \right) \cdot \left(1 - \frac{2\kappa}{(1+\kappa)n} \right)^t,$$

and in fact the proof establishes a linear decay of the expected suboptimality by the factor $1 - \frac{2\kappa}{(1+\kappa)n}$ in each single iteration. With a budget in place, the improvement is further reduced by the difference between the actual dual progress (the numerator in Eq. 6) and the progress in the non-budgeted case (the denominator in Eq. 6). By construction (see Lemma 1 and Eq. 6), this additive difference, when written as a multiplicative factor, amounts to $1 - E(w, \tilde{w})$ in the worst case. The worst case is reflected by taking the maximum over $i \in \{1, \dots, n\}$ in Eq. 6.

We interpret Theorem 1 as follows. The behavior of BSCA can be divided into an early and a late phase. For fixed weight degradation, the relative approximation error is small as long as the progress is sufficiently large, which is the case in early iterations. Then the algorithm is nearly unaffected by the budget constraint, and multiplicative progress at a fixed rate is achieved. This is consistent with non-budgeted SVM solvers [18].

Progress gradually decays when approaching the optimum. This increases the relative approximation error. At some point error and optimization progress cancel each other out and BSCA stalls. In fact, the theorem does not guarantee further progress for $E(w, \tilde{w}) \geq 1$. Due to $w^* \notin W_B$, the KKT violations do not decay to zero, and the algorithm approaches a limit distribution.[3] The precision to which the optimal SVM solution can be approximated is hence limited by the relative approximation error, or indirectly, by the weight degradation.

4.2 Budget Maintenance Rate

The rate at which budget maintenance is triggered can play a role, in particular if the procedure consumes a considerable share of the overall runtime. The following Lemma highlights that BSGD and BSCA can trigger budget maintenance at significantly different rates. For an algorithm A let

$$p_A = \lim_{T \to \infty} \mathbb{E} \left[\frac{1}{T} \cdot \left| \left\{ t \in \{1, \dots, T\} \,\middle|\, y_{i^{(t)}} \langle w^{(t-1)}, \phi(x_{i^{(t)}}) \rangle < 1 \right\} \right| \right]$$

denote the expected fraction of optimization steps in which the target margin is violated, in the limit $t \to \infty$ (if the limit exists). The following lemma establishes the fraction for primal SGD (Eq. (4)) and dual SCA (Eq. (3)), both without budget.

---

[3] BSCA does not converge to a unique point. It does not become clear from the analysis provided by [35] whether this is also the case for BSGD, or whether the decaying learning rate allows BSGD to converge to a local minimum.

Lemma 2. *Under the conditions (i)* $\alpha_i^* \in \{0, C\} \Rightarrow \frac{\partial D(\alpha^*)}{\partial \alpha_i} \neq 0$ *and (ii)* $\frac{\partial D(\alpha^{(t)})}{\partial \alpha_{i_t}} \neq 0$ *(excluding only a zero-set of cases) it holds* $p_{SGD} = \frac{1}{n} \sum_{i=1}^{n} \frac{\alpha_i^*}{C}$ *and* $p_{SCA} = \frac{1}{n} |\{i \,|\, 0 < \alpha_i^* < C\}|.$

Proof. The lemma considers the non-budgeted case, therefore the training problem is convex. Then the condition $\sum_{t=1}^{\infty} \frac{1}{t} = \infty$ for the learning rates ensures convergence $\alpha^{(t)} \to \alpha^*$ with SGD. This can happen only if the subtraction of $\alpha_i^{(t-1)}$ and the addition of nC with learning rate $\frac{1}{t}$ cancel out in the update Eq. (4) in the limit $t \to \infty$, in expectation. Formally speaking, we obtain

$$\lim_{T \to \infty} \mathbb{E}\left[\frac{1}{T} \sum_{t=1}^{T} \mathbf{1}_{\{i^{(t)}=i\}} \mathbf{1}_{\{y_{i(t)} \langle w^{(t-1)}, \phi(x_{i(t)}) \rangle < 1\}} nC - \alpha_i^{(t-1)} \right]$$
$$= 0 \quad \forall i \in \{1, \ldots, n\},$$

and hence $\lim_{T \to \infty} \mathbb{E}\left[\frac{1}{T} \sum_{t=1}^{T} \mathbf{1}_{\{i^{(t)}=i\}} \mathbf{1}_{\{y_{i(t)} \langle w^{(t-1)}, \phi(x_{i(t)}) \rangle < 1\}} \right] = \frac{\alpha_i^*}{nC}$. Summation over i completes the proof of the result for SGD.

In the dual algorithm, with condition (i) and the same argument as in [18] there exists an iteration t_0 so that for $t > t_0$ all variables fulfilling $\alpha_i^* \in \{0, C\}$ remain fixed: $\alpha_i^{(t)} = \alpha_i^{(t_0)}$, while all other variables remain free: $0 < \alpha_i^{(t)} < C$. Assumption (ii) ensures that all steps on free variables are non-zero and hence contribute $1/n$ to p_{SCA} in expectation, which yields $p_{SCA} = \frac{1}{n} |\{i \,|\, 0 < \alpha_i^* < C\}|$.

A point $(x_{i(t)}, y_{i(t)})$ that violates the target margin of one is added as a new support vector in BSGD as well as in BSCA. After the first B such steps, all further additions trigger budget maintenance. Hence Lemma 2 gives an asymptotic indication of the number of budget maintenance events. Strictly speaking, the result holds for the non-budgeted algorithms. It holds approximately if $\tilde{w} \approx w$, i.e., if the budget is not too small. This is very well confirmed experimentally. The different rates for primal and dual algorithm underline the quite different optimization behavior of the two algorithms: while (B)SGD keeps making non-trivial steps on all training points corresponding to $\alpha_i^* > 0$ (support vectors w.r.t. w^*), after a while the dual algorithm operates only on the free variables $0 < \alpha_i^* < C$.

5   Adaptive Coordinate Frequency on a Budget

In the budget stochastic coordinate ascent method, the active coordinate in Algorithm 1 has been chosen uniformly at random, in this section a more sophisticated implementation, namely the Adaptive Coordinate Frequencies (ACF) method described in [11] is embedded into the BSCA algorithm. ACF maintains a discrete probability distribution on the dual variables, hence adapts the coefficients with respect to the relative progress of the objective function, emphasizing coordinates that results in larger progress.

Algorithm 2. Adaptive Coordinate Frequencies for Budgeted Stochastic Coordinate Ascent (ACF-BSCA).

1: $\alpha \leftarrow 0$, $M \leftarrow \emptyset$
2: $a_i \leftarrow 0$, $p_i \sim U([p_{\min}, p_{\max}])$, $p_{\text{sum}} \leftarrow \sum_i p_i$, $\bar{r} \leftarrow 0$
3: **while** not happy **do**
4:     $J \leftarrow$ empty list
5:     **for** $i \in \{1, \ldots, n\}$ **do**
6:         $a_i \leftarrow a_i + n \cdot p_i/p_{\text{sum}}$
7:         add i $\lfloor a_i \rfloor$ times to J
8:         $a_i \leftarrow a_i - \lfloor a_i \rfloor$
9:     **end for**
10:    shuffle the list J
11:    **for** $i \in J$ **do**
12:        $\tilde{f}(x_i) = \sum_{(\beta, \tilde{x}) \in M} \beta k(x_i, \tilde{x})$
13:        $s \leftarrow \left(1 - y_i \tilde{f}(x_i)\right)/Q_{ii}$
14:        $\delta = [\alpha_i + s]_0^C - \alpha_i$
15:        **if** $\delta \neq 0$ **then**
16:            $\alpha_i \leftarrow \alpha_i + \delta$
17:            $M \leftarrow M \cup \{(\delta, x_i)\}$
18:            **if** $|M| > B$ **then**
19:                trigger budget maintenance, i.e., merge two support vectors
20:            **end if**
21:        **end if**
22:        $\Delta D = \frac{1}{2} Q_{ii} \cdot [s^2 - (s - \delta)^2]$
23:        $p_{\text{new}} \leftarrow \left[\exp\left(c \cdot (\Delta D/\bar{r} - 1)\right) \cdot p_i\right]_{p_{\min}}^{p_{\max}}$
24:        $p_{\text{sum}} \leftarrow p_{\text{sum}} + p_{\text{new}} - p_i$
25:        $p_i \leftarrow p_{\text{new}}$
26:        $\bar{r} \leftarrow (1 - \eta) \cdot \bar{r} + \eta \cdot \Delta D$
27:    **end for**
28: **end while**
29: **return** M
30:

ACF is an extension of the stochastic coordinate ascent algorithm. It does not treat all coordinates equally, but instead picks important coordinates more often than others. It thus makes faster progress to the optimum by modeling and adapting the relative frequencies for coordinate selection explicitly in the stochastic coordinate ascent algorithm. Adaptation of coordinate frequencies is driven by the progress made on the single-coordinate sub-problem. The optimization scheme results in enhanced performance and significant speed-ups in large scale machine learning problems.

The algorithm maintains a set of preference values p_i, one for each training point, as well as their sum $p_{\text{sum}} = \sum_{i=1}^{n} p_i$. It decomposes into two operations:

- efficient stratified sampling of a sequence of coordinates (roughly corresponding to one sweep over the data set), which follows the probabilities p_i/p_{sum},
- and updating of the preferences based on the observed progress.

The integration of this ACF algorithm into our BSCA approach is given in Algorithm 2, called ACF-BSCA.

The sampling step (lines 5 to 10) involves the probabilities p_i/p_{sum} as well as "accumulator" variables a_i which take care of rounding errors during stratified sampling. It produces a sequence J or indices. In contrast to uniform sampling from a discrete distribution, sampling of an index takes amortized constant time.

The update of the preferences in line 23 is based on the dual progress ΔD achieved in the current step with index i, or more precisely, with variable α_i. To this end the algorithm maintains an exponentially fading running average $\bar{r}$ of progress values. If ΔD exceeds $\bar{r}$ then the preference p_i is increased, while otherwise it is decreased. Over time, this scheme helps to balance the progress made by all coordinates [11].

6 Fast Randomized Merging

In this section we improve the merging step, which in itself can be rather costly. After only B updates of the model, each further update triggers the budget maintenance procedure. Its most costly operation is the identification of the ideal merge partner $(\beta_j, \tilde{x}_j)$ for the fixed first merge partner $(\beta_i, \tilde{x}_i)$, see Sect. 2.2. It requires the computation of the weight degradation for each of the B candidate indices. This step, although important for the proper function of the algorithm, is not in itself making optimization progress. We therefore aim to reduce its computational load.

This is achieved by not considering all B indices, but only a random subset of size independent of B. The proceeding is motivated by the Bayes Success-Run Theorem (based on the binomial distribution). It is a useful method for determining an appropriate risk-based sample size for process validations. In its simplest form, given N independent trials with a success probability of p, the theorem states that with confidence level C, the probability p can be bounded from below by the reliability

$$R = (1 - C)^{(1/N)}.$$

Solving for N yields $N = \frac{ln(1-C)}{ln(R)}$.

We are interested in obtaining a good merge partner j in a probably approximately correct (PAC) sense. We strive for j from the set of 5% best candidates, with 95% confidence. This is achieved by taking the best index from a random sample of size $N = 59 > 58.4 \approx \log(0.05)/\log(0.95)$ [30]. Compared to a typical budget size of $B = 500$ this represents a substantial saving of factor 8.5. We refer to this merging strategy as *random-59* in the following. In our implementation we restrict the search for merge partners to candidates of the same class, i.e., $\text{sign}(\beta_i) = \text{sign}(\beta_j)$.

Still, judging each merge partner requires solving a non-linear optimization problem for the placement of the merged point, see Sect. 2.2. For Gaussian kernels it can be reduced to a one-dimension problem, which is solved in [35] with golden section search (GSS). In this work we incorporate a much faster alternative method presented in [13], which is based on precomputing the problem combined with bilinear interpolation. We refer to that method as lookup-WD. The name refers to looking up the achievable weight degradation for each merge candidate.

7 Stopping Criterion

The BSGD solver does not feature a stopping criterion based on solution quality. In contrast, exact solvers like LIBSVM implement a stopping criterion based on the maximal KKT violation [2,17]. The approach fits particularly well with most violating pairs (MVP) working set selection [16]. In our setting, the maximal KKT violation is the maximal component of the dual gradient, with the maximum taken over dual variables α_i for which the bound constraint is not active or the gradient pushes the variable away from the constraint. For normalized kernels it is closely related to the length of the optimization step (see Eq. 3), which simply truncates the gradient component so that the solution remains feasible.

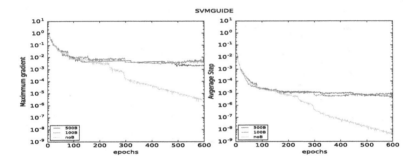

Fig. 1. Example of the maximum gradients curve (left) and average steps (right) with and without applying budget maintenance, i.e. merging strategy. The results are shown for SVMGUIDE data set.

Figure 1 shows that in the budgeted paradigm KKT violations are not suitable for designing a stopping criterion, since the budget constraint prevents the solution to converge to the unconstrained optimum, and it hence prevents the KKT violations from converging to zero. Instead they approach some positive value, which depends on the data set, on the budget B, and possibly on local optimum, since the training problem with budget is non-convex. This makes it impossible to define a meaningful threshold value a-priori. Even reliable detection of convergence is tricky, due to random fluctuations. However, the most compelling limitation of the approach is that the rather flat and noisy curves do not provide reliable hints at when the test error curves stabilize.

We aim to stop the solver when the test error starts to saturate and stops to oscillate. To achieve this goal we need a quantity that can be monitored online during the run of the algorithm, and the behavior of which is predictive of the above behavior, using a single uniform threshold across a wide range of data sets. In the following we propose a criterion with the afore-mentioned attributes.

Gradient values are usually used for stopping, however, Fig. 1 shows clearly that applying a budget affects convergence. Therefore we took a closer look at the individual sub-gradients, and we recorded statistics thereof over many optimization runs. It turned out that the minimum and maximum values of the sub-gradients yield smooth and nearly perfectly monotonic curves with easy-to-detect asymptotic behavior, see Fig. 2. Importantly, here the maximum is taken over the actual optimization steps over one epoch, in contrast to all possible steps at a single moment in time, as considered by the KKT violations. Hence, the quantity is related to the KKT conditions, but it is a heuristic. A related heuristic is applied in LIBLINEAR [10], which accumulates KKT violations over one epoch.

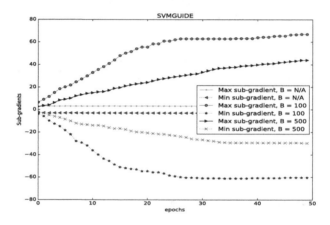

Fig. 2. Min./max. sub-gradients curves of the dual decomposition algorithm on a budget for the SVMGUIDE data set.

The main advantage of the curves in Fig. 2 over those in Fig. 1 is smoothness, which greatly simplifies reliable detection of convergence. At the same time we observe empirically that the test error tends to stabilize roughly at the point where the curves in Fig. 2 start to saturate. However, the absolute values vary a lot across data sets. In order to obtain a scale-independent measure we monitor the *logarithm* of the difference between maximum and minimum gradient, see Fig. 3. Even the finite difference curves estimating the slope are rather smooth. Their uniform scale allows us to device a uniform threshold value. In our experience, thresholds between 0.001 and 0.01 work well in practice—see also the experiments in the next section.

8   Experiments

The main goal of our experiments is to demonstrate the massive speed-ups achievable with the toolkit of methods presented in the previous sections. We start by analyzing each method in isolation, and finally combine the different pieces to a single powerful solver well suited for large-scale non-linear SVM training on a budget.

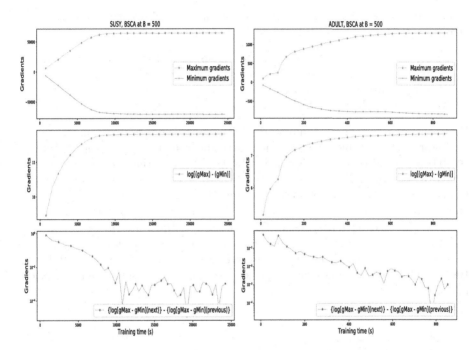

Fig. 3. Min./max. sub-gradients curves of the dual decomposition algorithm on a budget. The results are shown for SUSY and ADULT data sets on a budget of $B = 500$.

Table 1. Data sets used in this study, hyperparameter settings, test accuracy, number of support vectors, and training time of the full SVM model (using LIBSVM).

Data set	Size	Features	C	γ	Accuracy	#SVs	Training time
SUSY	4,500,000	18	2^5	2^{-7}	80.02%	2,052,555	504 h 25 m 38 s
COVTYPE	581,012	54	2^7	2^{-3}	75.88%	187,626	10 h 5 m 7 s
KDDCUP08	102,294	117	2^5	2^{-7}	99.45%	1,806	118.456 s
COD-RNA	59,535	8	2^5	2^{-3}	96.33%	9,120	53.951 s
IJCNN	49,990	22	2^5	2^1	98.77%	2,477	46.914 s
ADULT	32,561	123	2^5	2^{-7}	84.82%	11,399	97.152 s

The experiments are based on the data sets listed in Table 1, which provides descriptive statistics of the data and of the optimal SVM solution. The data sets cover a wide range of problem sizes, with up to 4.5 million for the SUSY problem, which took three weeks to train with LIBSVM. Unless stated otherwise, we used a budget size of $B = 500$.

8.1   Primal Vs Dual Training

In our first block of experiments we conduct a comparison of the novel BSCA algorithm to state-of-the-art methods, in particular to BSGD.

1. We analyze learning curves, in particular speed and stability of learning, in terms of prediction accuracy, and primal as well as dual objective value.
2. We investigate the impact of the budget size on optimization behavior.
3. We compare BSCA to a large number of state-of-the-art budget methods.

We present a representative summary of the comparison results from [28].

Figure 4 shows typical convergence curves of primal objective (optimized by BSGD) and dual objective (optimized by BSCA), and the corresponding evolution of the test accuracy. We see that the dual BSCA solver clearly outperforms primal BSGD. The dual objective function values are found to be smooth and monotonic, which is not the case for the primal. BSCA generally oscillates less and stabilizes faster, while BSGD remains unstable and has the possibility of delivering low-quality solution for much longer time. Similar behavior is observed for the accuracy curve. It converges faster for the dual solver, while the BSGD suffers from a long series of peaks corresponding to significantly reduced performance.

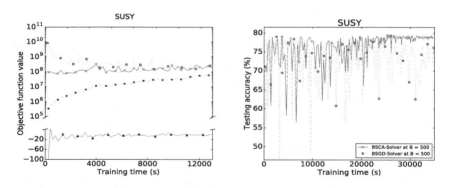

Fig. 4. Objective function and testing accuracy curves of BSGD and BSCA for the SUSY data set.

The next experiment investigates the impact of budget size on the optimization and learning behavior. For this purpose, the budget size is reduces to $B = 200$. It is understood that the iteration complexity is reduced significantly, however, it remains to be seen whether this effect translates into improved learning behavior. Figure 5 reveals that this can indeed be the case, e.g., for the SUSY data, where a significant speed-up is achieved without compromising accuracy. However, for the IJCNN data the reduction in budget size is fatal, and the larger model achieves better accuracy in much less time. Finally, care needs to be taken to avoid too small budgets in BSCA, since it can result is drift effects when the budgeted representation $\tilde{w}$ does not properly reflect the full model w any more.

An extensive set of experiments was conducted to examine the efficiency of the BSCA algorithm in comparison to 11 state-of-the-art online kernel learning approaches considered in [20]. The results are shown in Fig. 6. The results demonstrate the effectiveness of BSCA in terms of fast training and high accuracy. They validate the effectiveness and efficiency of BSCA algorithm for large-scale kernel learning.

8.2   The Effect of Adaptive Coordinate Frequencies

In order to assess the effect of adaptive vs. uniform coordinate sampling we compare ACF-BSCA to its natural baseline, the BSCA algorithm with uniform coordinate selection as described in Sect. 3. In these experiments, the number of epochs is fixed to 50 for all data sets, with the exception of COVTYPE, where 100 epochs were used to achieve stable performance. We investigate the effect of the ACF method on training time, predictive accuracy, primal and dual objective, and their stability over the course of training.

Detailed convergence plots for primal and dual objective value are displayed in Fig. 7. Overall, ACF-BSCA stabilizes faster than BSCA. In most cases, the dual objective value rises faster (with the exception of COVTYPE, where there is no effect), which underlines the superiority of ACF-BSCA as an optimization method. With ACF an epoch take slightly longer than without. This is due to two effects: First, the length of an epoch is not fixed in Algorithm 2, and second, the success of the ACF method in avoiding useless zero-steps on non-support-vectors results in more progress, which also results in more frequent merging.

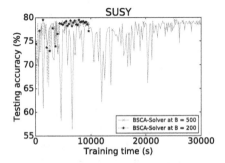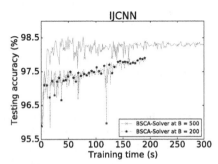

Fig. 5. Test accuracy results for BSCA and BSGD at budgets of 200 and 500 for the SUSY and IJCNN data sets.

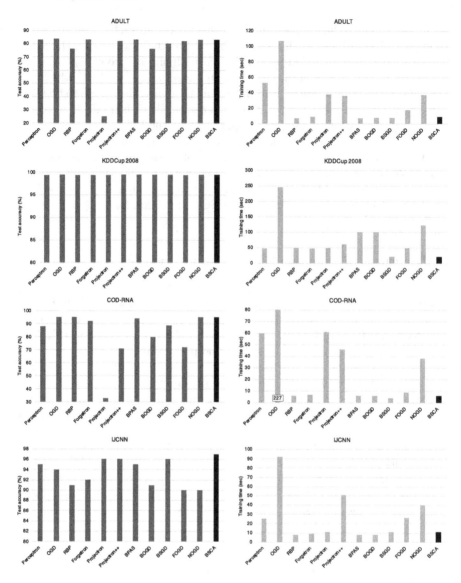

Fig. 6. Training time (seconds) and testing accuracy (%) for BSCA and all approaches implemented by [20].

Figure 8 shows the corresponding evolution of the test error, including its variability across 10 independent runs of both algorithms. ACF-BSCA starts with mixed results in the first epochs. This is because ACF starts with random preferences, and learning the preferences takes some time. After a few epochs, the learning curves stabilize significantly for SUSY, COD-RNA and ADULT. However, ACF does not always work well. In th COVTYPE problem, where the

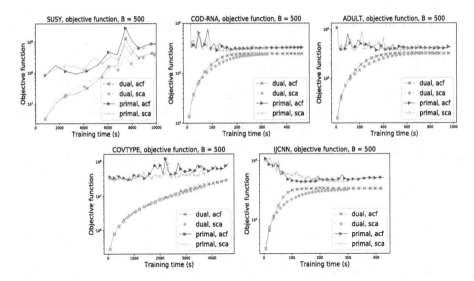

Fig. 7. Primal and dual objective function curves of BSCA and ACF-BSCA solvers with budget size $B = 500$ for data set SUSY, COVTYPE, COD-RNA, IJCNN, and ADULT.

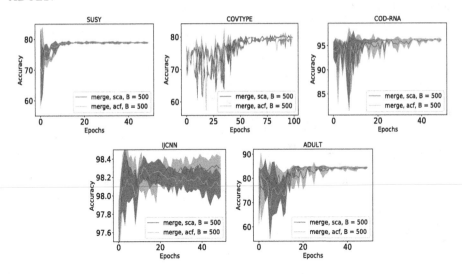

Fig. 8. Test accuracy of BSCA and ACF-BSCA at a budget of $B = 500$. The plots indicate mean and standard deviation of the test accuracy over 10 independent runs for the data sets SUSY, COVTYPE, COD-RNA, IJCNN, and ADULT.

algorithm did not show any advantage in terms of optimization, the results are mixed. For IJCNN the solution even degrades in the late phase. This happens due to too large weight degradations, and it can be avoided by increasing the budget.

8.3   The Effect of Random-59 Sampling and Lookup-WD

In this section we investigate the impact of combining our sampling scheme *random-59* and the lookup algorithm [13].

Figure 9 shows the test error for BSGD and BSCA models with full merging, and BSCA with random-59 merging. We have already discussed the superiority of BSCA over BSGD, which is confirmed once more. Here we focus on the speed-up, i.e., the saving in terms of training time of random-59 over full merging. We observe savings of up to 25%. They are particularly pronounced for the large data sets COVTYPE and SUSY. Importantly, test performance does not suffer from the sampling heuristic. Generally speaking, the random-59 sampling method outperforms full merging without affecting the test accuracy.

In the next step we also include the lookup-WD method into the merging procedure. This leaves us with four combinations: full merging vs. random-59 combined with GSS vs. lookup-WD. We also compare to BSGD for reference. All methods found SVM models of comparable accuracy; in fact the differences were within one standard deviation of the variability between different runs. The runtimes are shown in Fig. 10. It becomes clear that in particular the random-59 heuristic is highly effective, while lookup-WD adds another speed-up on top. In combination, the techniques make up for the slightly larger iteration cost of BSCA over BSGD.[4]

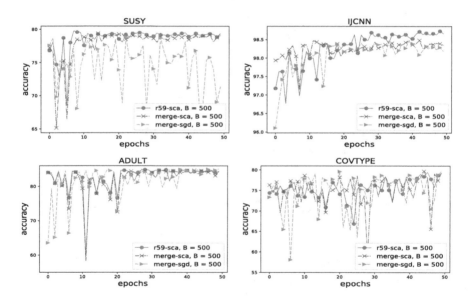

Fig. 9. Test accuracy curves for BSCA with random-59, compared to BSGD and BSCA with full merging as baselines, on the SUSY, IJCNN, ADULT, and COVTYPE data sets.

---

[4] However, note that iteration time is not a suitable measure of learning speed, since BSCA makes significantly more progress per iteration and stabilizes faster than BSGD.

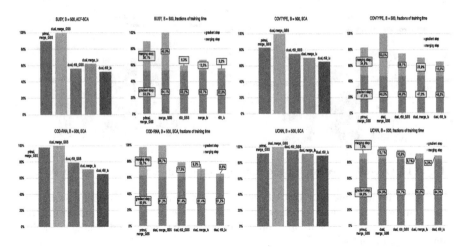

Fig. 10. Relative training time of five algorithms, normalized to the training time of plain BSCA with full merging and golden section search (the 100% mark). Training times from left to right: BSGD, BSCA (both with full merging and GSS), BSCA with random-59 and GSS, BSCA with full merging and lookup-WD, and BSCA with the combination of random-59 and lookup-WD. The pink bars (bottom part of the right part of each figure) represent the gradient/margin computation, while the green bars (top part) represent budget maintenance.

We go one step further and provide a breakdown of the total training time into 1. margin computation (or gradient computation) and 2. merging step, see Fig. 10. For all data sets, the time spent in the gradient step is similar, while the time spent on budget maintenance differs significantly between the methods. It is expected that Lookup-WD is faster than GSS (see also [13]) and that random-59 is faster than full merging. The expectation is clearly confirmed by the experimental data.

The greatest saving we gain when applying random-59 and lookup-WD is about 45.0% of the total training time for the large SUSY data set. For other data sets like IJCNN the speed-ups are less pronounced. In general, the potential for savings depends primarily on the rate at which merging occurs. However, it is worth emphasizing that the proposed method is never worse than the baseline.

8.4 Adaptive Stopping Criterion

The advantage of an adaptive stopping criterion is that it can stop the solver when the solution quality is "just right". Here we investigate whether our proposed criterion has this quality, which must be achieved with a single threshold value across a wide range of data sets. Furthermore, the threshold must be set to a value that guarantees sufficient precision without wasting computation time.

The widely agreed-upon criterion of LIBSVM is to accept a maximal KKT violation of $\varepsilon = 10^{-3}$. A lower precision of 10^{-2} or even 10^{-1} (depending on the mode) is used by default in LIBLINEAR.

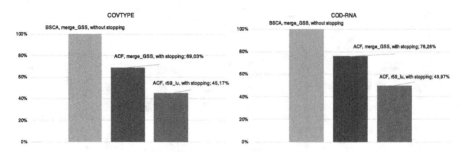

Fig. 11. Overall speed-up for the combination of random-59 and lookup-WD.

Table 2. Training time and test accuracy on different data sets achieved by sub-gradient stopping criterion. Training time and test accuracy are tested at different thresholds, ACF-BSCA results are compared to the BSCA method. Budget = 500 for all data sets and methods.

Data set	Method		Stopping at threshold ($\epsilon < 0.01$)	Stopping at threshold ($\epsilon < 0.001$)
SUSY	BSCA	Training time	7939.71 s	9849.54 s
		Testing accuracy	$(78.29 \pm 0.85)\%$	$(79.39 \pm 0.19)\%$
	ACF-BSCA	Training time	8001.39 s	9650,877 s
		Testing accuracy	$(77.75 \pm 0.48)\%$	$(78.40 \pm 0.45)\%$
COVTYPE	BSCA	Training time	3634.71 s	6719.23 s
		Testing accuracy	$(76.31 \pm 1.02)\%$	$(80.20 \pm 0.28)\%$
	ACF-BSCA	Training time	2905.14 s	4638.35 s
		Testing accuracy	$(74.23 \pm 0.38)\%$	$(77.94 \pm 1.04)\%$
COD-RNA	BSCA	Training time	152.38 s	198.36 s
		Testing accuracy	$(96.27 \pm 1.05)\%$	$(95.94 \pm 0.39)\%$
	ACF-BSCA	Training time	131.09 s	151.26 s
		Testing accuracy	$(95.97 \pm 2.94)\%$	$(95.54 \pm 1.34)\%$
IJCNN	BSCA	Training time	125.57 s	143.17 s
		Testing accuracy	$(98.21 \pm 0.21)\%$	$(98.39 \pm 0.12)\%$
	ACF-BSCA	Training time	94.01 s	126.60 s
		Testing accuracy	$(98.20 \pm 0.09)\%$	$(98.35 \pm 0.14)\%$
ADULT	BSCA	Training time	322.51 s	576.74 s
		Testing accuracy	$(83.19 \pm 0.48)\%$	$(84.02 \pm 1.31)\%$
	ACF-BSCA	Training time	382.56 s	479.08 s
		Testing accuracy	$(84.12 \pm 1.23)\%$	$(84.29 \pm 1.24)\%$

Table 2 presents training times and test accuracies of BSCA with and without ACF with our stopping criterion for two different candidate thresholds 0.01 and 0.001. First of all, it turns out that the training time does not depend critically on the value of ε, with a maximal difference of about factor two for the COVTYPE data, and differences of roughly 25% to 50% for the other cases. The accuracy values indicate that for some problems the lower solution precision encoded by $\varepsilon = 0.01$ suffices, while in particular the COVTYPE problem profits from a higher precision, corresponding to $\varepsilon = 0.001$. Therefore we generally recommend the threshold value of $\varepsilon = 0.001$, which is also used by other kernelized SVM training methods, as a robust default.

8.5   Overall Speed-Up

The new stopping criterion allows for a more direct comparison of our various contributions. We have shown the vast superiority of BSCA over the BSGD approach. However, even within this class of solvers significant speed-ups are achievable. For the COVTYPE data, the ACF approach saves about 31% of the training time (see Table 2). The combination of random-59 and lookup-WD saves another 31%. Taken together the speed-ups multiply to more than 50%. The results are presented in Fig. 11. For other data sets the effect is less pronounced, but still relevant.

9   Conclusion

In this paper we have designed a fast dual algorithm for non-linear SVM training on a budget. It combines many different components into a single solver: We replace primal BSGD with dual BSCA, incorporate adaptive coordinate frequencies (ACF) for improved working variable selection, speed up merging through a randomization heuristic and a fast lookup, and introduce a robust and reliable stopping criterion.

The BSCA algorithm greatly outperform the primal BSGD method as an optimization algorithm (in terms of primal and dual objective function) and as a learning algorithm (in terms of predictive accuracy). The ACF method offers a further speedup, in particular for the dual objective.

We have demonstrated that our *random-59* search heuristic for merge candidates can speed up budget maintenance by a significant factor. Furthermore, we combine this technique with a fast lookup. With these techniques, we reduce the budget maintenance overhead by about factor 10, to only a few percent of the overall runtime.

Finally, we have designed a robust and reliable stopping criterion for dual training with budget. It is related to the KKT-based stopping criterion of exact solvers, but stabilized through aggregation over a full epoch. By performing a logarithmic transformation we obtain an extremely robust criterion that works well and reliable across a wise range of data sets.

We have demonstrated empirically that all of the above techniques contribute to a faster and/or more stable training process. Compared to the state-of-the-art BSGD algorithm, the combined improvements in speed and stability are tremendous. They mark significant progress, enabling truly large-scale kernel learning. We can clearly recommend our methods, BSCA and ACF-BSCA, as plug-in replacements for primal methods. It is rather straightforward to extend our algorithms to other kernel machines with box-constrained dual problems, like regression and some multi-class classification schemes.

Acknowledgments. We acknowledge support by the Deutsche Forschungsgemeinschaft (DFG) through grant GL 839/3-1.

References

1. Bordes, A., Bottou, L., Gallinari, P., Weston, J.: Solving multiclass support vector machines with LaRank. In: International Conference on Machine Learning (2007)
2. Bottou, L., Lin, C.J.: Support vector machine solvers (2006)
3. Burges, C.J.: Simplified support vector decision rules. In: ICML, pp. 71–77 (1996)
4. Cavallanti, G., Cesa-Bianchi, N., Gentile, C.: Tracking the best hyperplane with a simple budget perceptron. Mach. Learn. **69**(2–3), 143–167 (2007)
5. Chang, C.C., Lin, C.J.: LIBSVM: a library for support vector machines. ACM Trans. Intell. Syst. Technol. **2**(3), 27 (2011)
6. Chen, P., Fan, R., Lin, C.: A study on SMO-type decomposition methods for support vector machines. IEEE Trans. Neural Netw. **17**(4), 893–908 (2006)
7. Cortes, C., Vapnik, V.: Support-vector networks. Mach. Learn. **20**(3), 273–297 (1995)
8. Crammer, K., Kandola, J., Singer, Y.: Online classification on a budget. In: Advances in Neural Information Processing Systems 16, pp. 225–232. MIT Press (2004)
9. Dekel, O., Shalev-Shwartz, S., Singer, Y.: The forgetron: a kernel-based perceptron on a budget. SIAM J. Comput. **37**(5), 1342–1372 (2008)
10. Fan, R.E., Chang, K.W., Hsieh, C.J., Wang, X.R., Lin, C.J.: LIBLINEAR: a library for large linear classification. J. Mach. Learn. Res. **9**, 1871–1874 (2008)
11. Glasmachers, T., Dogan, Ü.: Accelerated coordinate descent with adaptive coordinate frequencies. In: ACML. JMLR Workshop and Conference Proceedings, vol. 29, pp. 72–86. JMLR.org (2013)
12. Glasmachers, T., Igel, C.: Maximum-gain working set selection for SVMs. J. Mach. Learn. Res. **7**, 1437–1466 (2006)
13. Glasmachers, T., Qaadan, S.: Speeding up budgeted stochastic gradient descent SVM training with precomputed golden section search (2018)
14. Joachims, T.: Text categorization with support vector machines: learning with many relevant features. In: Nédellec, C., Rouveirol, C. (eds.) ECML 1998. LNCS, vol. 1398, pp. 137–142. Springer, Heidelberg (1998). https://doi.org/10.1007/BFb0026683
15. Joachims, T.: Making large-scale support vector machine learning practical. In: Advances in Kernel Methods, pp. 169–184. MIT Press, Cambridge (1999)
16. Keerthi, S., Shevade, S., Bhattacharyya, C., Murthy, K.: Improvements to Platt's SMO algorithm for SVM classifier design. Technical report (1999)
17. Keerthi, S., Gilbert, E.: Convergence of a generalized SMO algorithm for SVM classifier design. Mach. Learn. **46**(1), 351–360 (2002)
18. Lin, C.J.: On the convergence of the decomposition method for support vector machines. IEEE Trans. Neural Netw. **12**(6), 1288–1298 (2001)
19. List, N., Simon, H.U.: General polynomial time decomposition algorithms. In: Auer, P., Meir, R. (eds.) COLT 2005. LNCS (LNAI), vol. 3559, pp. 308–322. Springer, Heidelberg (2005). https://doi.org/10.1007/11503415_21

20. Lu, J., Hoi, S.C., Wang, J., Zhao, P., Liu, Z.Y.: Large scale online kernel learning. J. Mach. Learn. Res. **17**(47), 1–43 (2016)
21. Ma, S., Belkin, M.: Diving into the shallows: a computational perspective on large-scale shallow learning. In: Neural Information Processing Systems (NIPS), pp. 3778–3787 (2017)
22. Mohri, M., Rostamizadeh, A., Talwalkar, A.: Foundations of Machine Learning. MIT press, Cambridge (2012)
23. Nesterov, Y.: Efficiency of coordinate descent methods on huge-scale optimization problems. SIAM J. Optim. **22**(2), 341–362 (2012)
24. Nguyen, D., Ho, T.: An efficient method for simplifying support vector machines. In: International Conference on Machine Learning, pp. 617–624. ACM (2005)
25. Orabona, F., Keshet, J., Caputo, B.: Bounded kernel-based online learning. J. Mach. Learn. Res. **10**, 2643–2666 (2009)
26. Osuna, E., Freund, R., Girosi, F.: An improved training algorithm of support vector machines. In: Neural Networks for Signal Processing VII, pp. 276–285, October 1997
27. Platt, J.C.: Sequential minimal optimization: a fast algorithm for training support vector machines. Technical report, Advances in kernel methods - support vector learning (1998)
28. Qaadan, S., Schüler, M., Glasmachers, T.: Dual SVM training on a budget. In: Proceedings of the 8th International Conference on Pattern Recognition Applications and Methods. SCITEPRESS - Science and Technology Publications (2019)
29. Schölkopf, B., et al.: Input space versus feature space in kernel-based methods. IEEE Trans. Neural Netw. **10**(5), 1000–1017 (1999)
30. Schölkopf, B., Smola, A.: Learning with Kernels: Support Vector Machines, Regularization, Optimization, and Beyond. MIT Press, Cambridge (2001)
31. Shalev-Shwartz, S., Singer, Y., Srebro, N.: Pegasos: primal estimated sub-gradient solver for SVM. In: Proceedings of the 24th International Conference on Machine Learning, pp. 807–814 (2007)
32. Steinwart, I.: Sparseness of support vector machines. J. Mach. Learn. Res. **4**, 1071–1105 (2003)
33. Steinwart, I., Hush, D., Scovel, C.: Training SVMs without offset. J. Mach. Learn. Res. **12**(Jan), 141–202 (2011)
34. Wang, Z., Vucetic, S.: Tighter perceptron with improved dual use of cached data for model representation and validation. In: IJCNN, pp. 3297–3302. IEEE Computer Society (2009)
35. Wang, Z., Crammer, K., Vucetic, S.: Breaking the curse of kernelization: budgeted stochastic gradient descent for large-scale SVM training. J. Mach. Learn. Res. **13**(1), 3103–3131 (2012)
36. Weston, J., Bordes, A., Bottou, L., et al.: Online (and offline) on an even tighter budget. In: AISTATS. Citeseer (2005)
37. Weston, J., Mukherjee, S., Chapelle, O., Pontil, M., Poggio, T., Vapnik, V.: Feature selection for SVMs. In: Leen, T.K., Dietterich, T.G., Tresp, V. (eds.) Advances in Neural Information Processing Systems 13, pp. 668–674 (2001)

Attributes for Understanding Groups of Binary Data

Arthur Chambon[1], Frédéric Lardeux[1(✉)], Frédéric Saubion[1(✉)], and Tristan Boureau[2(✉)]

[1] LERIA, Université d'Angers, Angers, France
{arthur.chambon,frederic.lardeux,frederic.saubion}@univ-angers.fr
[2] UMR1345 IRHS, Université d'Angers, Angers, France
tristan.boureau@univ-angers.fr

Abstract. In this paper, we are interested in determining relevant attributes for multi-class discrimination of binary data. Given a set of observations described by the presence or absence of a set of attributes and divided into groups, we seek to determine a subset of attributes to explain and describe these groups. A pattern is set of Boolean values that are shared by many observations in a given group. Thanks to a new pattern computation algorithm, we present an approach to optimize the choice of the important attributes. Using real biological instances, we compare our results with two other different approaches and discuss the difference in information obtained by each.

Keywords: Logical analysis of data · Multiple characterizations · Diagnostic test

1 Introduction

Logical analysis of data (LAD) [13] is a methodology based on the determination of combinations of binary values, called patterns, discriminating a subset of a group of data accurately. This is an original alternative to conventional data analysis, which consists of statistical methods of data analysis, but whose purpose is also to study a dataset for descriptive purposes and/or predictive.

Logical analysis of data consists of studying two groups of Boolean data (called observations) P and N (respectively positive and negative observations). These observations are described by the presence or absence of attributes (where the set of attributes is denoted $\mathcal{A}$). The objective of the logical data analysis is to determine a subset of $\mathcal{A}$ sufficient to discriminate observations from positive ad negative group in order to deduce logical rules/formulas explaining the dataset.

Logical analysis of data (LAD) combines concepts from partially defined Boolean functions and optimization in order to characterize such sets of data. Many applications have been pointed out, e.g., for diagnosis purposes when a physician wants to identify common symptoms that are shared by a group of patients suffering from similar diseases.

© Springer Nature Switzerland AG 2020
M. De Marsico et al. (Eds.): ICPRAM 2019, LNCS 11996, pp. 48–70, 2020.
https://doi.org/10.1007/978-3-030-40014-9_3

Note that classic classification methods used in machine learning (for instance clustering algorithms) compute and assign clusters/groups to incoming data. Our goal is merely to compute an explicit justification of the data. Our problem is also different from frequent itemsets computation in the context of data mining problems. In our context, the groups of data are provided by experts based on specific knowledge. Therefore we are definitely not dealing with a classification problem.

As an example, let us consider a set of 8 Boolean attributes (labels are α to κ) and 7 observations. These data have been put into two groups P and N.

Observations	Groups	Attributes							
		α	β	γ	δ	ϵ	τ	η	κ
1	P	1	1	0	0	1	0	0	1
2		0	1	0	1	0	1	0	1
3		1	1	0	0	1	0	0	0
4		0	0	0	1	0	0	0	1
5	N	1	0	1	0	0	1	0	0
6		0	1	0	1	1	0	1	1
7		1	0	0	1	1	1	0	1
8		1	1	1	0	0	1	1	0

In LAD methodology, a key concept is to determine patterns. A pattern in a subset of similar attributes values observed in several observations of a group. In our example, $\alpha = 0$ and $\beta = 1$ constitutes a pattern. This pattern is present in observations 1 and 3 of group P. But note that no observation of N has this pattern. Such a pattern constitutes a partial explanation of the properties of observations of group P. Since there are a multitude of patterns, it is necessary to decide which compromise must be reached between their size and the coverage they provide. Some properties have been exposed to determine the most relevant according to their size or coverage. In prime patterns, the number of attributes cannot be reduced unless they loose their pattern property. Strong patterns are patterns covering the maximal number of observations in considered group P.

Prime patterns are related to simplicity requirement (considering the number of attributes). Strong patterns are related to an evidential preference since a larger cover is preferred (the reader may read [12] for a more details on LAD).

Let us remark that attributes δ and ϵ can also used to generate a Boolean formula $\phi \equiv (\delta \wedge \neg\epsilon) \vee (\neg\delta \wedge \epsilon)$. This formula is true for P (if P is considered as a set of Boolean assignments and attributes are Boolean variables) and the formula is false for N. This formula involves two patterns, $\delta = 1$ and $\epsilon = 0$ and cover observation 2 and 4 and the second being $\delta = 0$ and $\epsilon = 1$ and cover observation 1 and 3.

ϕ is formulated in disjunctive normal form (DNF). Such a formula can be useful for experts and users. It is possible to minimize the number of attributes which may ease its their practical use in diagnosis routines. It is possible to minimize the size of the formula, which may improve its interpretation. Such techniques have been studied in [11] and even extended to more than two groups of data.

Motivations. In this paper, our main goal is to study different techniques to select relevant attributes to help the user to better understand groups of binary data and eventually to identify relevant properties of data. Hence, we consider the two previously described approaches: patterns and minimal sets of attributes. We also aim to compare these methods with a feature selection technique (note that feature selection methods are commonly used to improve classification algorithms by identifying relevant attributes, but may also be useful for data visualization).

Contributions. In [10] we proposed a new algorithm to compute all prime patterns and there coverage.

In this work, we propose (1) a new way to compute all Boolean formula ϕ with minimum size, (2) to compare the attributes that are selected by the two above-mentioned logical characterization techniques, (3) to compare these approaches with a feature selection method and (4) to study if the groups defined by experts could be explained by classification using the selected attributes. At last, we perform experiments on different real benchmarks issued from biology.

2   Logical Methods for Binary Data Characterization Problems

Logical Analysis of Data (LAD) [3,6,12,17] aims at characterizing groups of binary data. Given two groups (positive an negative groups) of observations (i.e., Boolean vectors), the purpose is to compute a justification/explanation of the membership of data to the groups. A justification is a formula that is satisfied by observations, considered as Boolean assignments, of one group while being falsified by the observations of the other group. The attributes appearing in the formula, discriminate thus the groups. LAD use concepts from partially defined Boolean functions [22] and focuses on the notion of pattern that corresponds to a subset of attributes whose values are similar for several observations in the positive group. This approach helps the user to identify common characteristics of these observations. LAD has been applied to many practical application domains: biology and medicine [1,2,20], engineering [4], transportation [14]. The Multiple Characterization Problem (MCP) [11] is an extension of the LAD methodology that considers simultaneously several groups of observations. MCP consists in minimizing the number of attributes that are necessary to discriminate mutually several groups of observations.

From these previous works, we propose a unified presentation of these possible characterizations of groups of binary data.

2.1 Basic Principles: Boolean Functions

A Boolean function f is a mapping $f : \mathbb{B}^n \mapsto \mathbb{B}$, where $\mathbb{B}$ is the set $\{0, 1\}$, n is the number of variables. A vector $x \in \mathbb{B}^n$ is a *true vector* (resp. *false vector*) of the Boolean function f if $f(x) = 1$ (resp. $f(x) = 0$). $T(f)$ (resp. $F(f)$) is the set of *true vectors* (resp. *false vectors*) of a Boolean function f.

A partially defined Boolean function (pdBf) on $\mathbb{B}^n$ is a pair (P, N) such that $P, N \subseteq \mathbb{B}^n$ and $P \cap N = \emptyset$. P is thus the set of positive vectors, and N the set of negative vectors of the pdBf (P, N).

We present the notion of term [5,17] that generalizes the notion of partially defined Boolean function. A term is a Boolean function t_{σ^+, σ^-} whose true set $T(t_{\sigma^+, \sigma^-})$ is of the form: $T(t_{\sigma^+, \sigma^-}) = \{x \in \mathbb{B}^n | x_i = 1 \; \forall i \in \sigma^+, \; x_j = 0 \; \forall j \in \sigma^-\}$ for some set $\sigma^+, \sigma^- \subseteq \{1, 2, ..., n\}$, $\sigma^+ \cap \sigma^- = \emptyset$. A term t_{σ^+, σ^-} can be represented by a Boolean formula of the form: $t_{\sigma^+, \sigma^-}(x) = (\bigwedge_{i \in \sigma^+} x_i) \wedge (\bigwedge_{j \in \sigma^-} \neg x_j)$.

2.2 Formulation of the Problem

Let us introduce the notion of Binary Data Characterization Problem (BDCP).

Definition 1. *An instance of the Binary Data Characterization Problem is a tuple $(\Omega, \mathcal{A}, D, G)$ defined by a set of observations $\Omega \subseteq \mathbb{B}^{|\mathcal{A}|}$ of Boolean vectors built on a set $\mathcal{A}$ of Boolean attributes. The observations are recorded in a Boolean matrix $D_{|\Omega| \times |\mathcal{A}|}$. A function $G : \Omega \to \{P, N\}$ assigns a group $G(o)$ to the observation $o \in \Omega$.*

The matrix D is defined as:

- *the value $D[o, a]$ represents the presence/absence of the attribute a in the observation o.*
- *a line $D[o, .]$ represents the Boolean vector of presence/absence of the different attributes in the observation o.*
- *a column $D[., a]$ represents the Boolean vector of presence/absence of the attribute a in all the observations.*

Given a subset $A \subset \mathcal{A}$, D^A is a matrix reduced to the attributes of A.

As already mentioned, two possible notions of solution are considered:

- computing minimal sets of attributes that discriminate the groups,
- computing patterns that are shared by observations of the positive group.

2.3 Minimal Sets of Attributes

Given a BDCP instance $(\Omega, \mathcal{A}, D, G)$ the purpose here is to find a subset of attributes $S \subseteq \mathcal{A}$ such that two observations from two different groups are always different on at least one attribute in S.

Definition 2. *Given an instance* $(\Omega, \mathcal{A}, D, G)$, *a subset of attributes* $S \subseteq \mathcal{A}$ *is a solution iff* $\forall (o, o') \in \Omega^2, G(o) \neq G(o') \Rightarrow D^S[o, .] \neq D^S[o', .]$. *In this case, the matrix* D^S *is called a solution matrix.*

An instance may have several solutions of different sizes. It is therefore important to define an ordering on solutions in order to compare and classify them. In particular, for a given solution S, adding an attribute generates a new solution $S' \supset S$. In this case we say that S' is dominated by S. We can also compute solutions of minimal size with regards to the attributes they involve.

Definition 3. *A solution* $S \subseteq \mathcal{A}$ *is non-dominated iff* $\forall s \in S, \exists (o, o') \in \Omega^2$ *s.t.* $G(o) \neq G(o')$ *and* $D^{S \setminus \{s\}}[o, .] = D^{S \setminus \{s\}}[o', .]$. *A solution* $S \subseteq \mathcal{A}$ *is minimal iff* $\nexists S' \subseteq \mathcal{A}$ *with* $|S'| < |S|$ *s.t.* S' *is a solution.*

According to previous works [11], given an instance $(\Omega, \mathcal{A}, D, G)$, the BDCP can be formulated as the following 0/1 linear program:

$$min : \sum_{i=1}^{|\mathcal{A}|} y_i$$

$$s.t. :$$

$$C \cdot Y^t \geqslant \mathbb{1}^t$$

$$Y \in \{0, 1\}^{|\mathcal{A}|}, Y = [y_1, ..., y_{|\mathcal{A}|}]$$

where Y is a Boolean vector that encodes the presence/absence of the set of attributes in the solution. C is a matrix that defines the constraints that must be satisfied in order to ensure that Y is a solution. Let us denote Θ the set of all pairs $(o, o') \in \Omega^2$ such that $G(o) \neq G(o')$. For each pair of observations (o, o') that does not belong to the same group, one must insure that the value of at least one attribute differs from o to o'. This will be insured by the inequality constraint involving the $\mathbb{1}$ vector (here a vector of dimension $|\Theta|$ that contains only values equal to 1). The minimization objective function insures that we aim to find a minimal solution.

More formally, C is a Boolean matrix of size $|\Theta| \times |\mathcal{A}|$ defined as:

- Each line is numbered by a couple of observations $(o, o') \in \Omega^2$ such as $G(o) \neq G(o')$ $((o, o') \in \Theta)$.
- Each column represents an attribute.
- $C[(o, o'), a] = 1$ if $D[o, a] \neq D[o', a]$, $C[(o, o'), a] = 0$ otherwise.
- We denote $C[(o, o'), .]$ the Boolean vector representing the differences between observations o and o' on each attribute. This Boolean vector is called constraint since one attribute a such $C[(o, o'), a] = 1$ must be selected in order to insure that no identical observations can be found in different groups.

Two algorithms have been proposed to compute solutions [9]:

– NDS (Non Dominated Solutions) that computes the set of all non-dominated solutions.

This algorithm find all $Y \in \{0,1\}^{|\mathcal{A}|}$ of the linear program above such as $C \cdot Y^t \geq 1^t$.

– MWNV (Merging with negative variables) that computes all minimal non dominated solutions.

This algorithm finds all the solutions of the linear program above.

Note that the computation of all minimal non dominated solutions is related to the Min Set Cover problem and the Hitting Set problem [19].

2.4   Patterns

Let us consider $P = \{o \in \Omega | G(o) = P\}$ the group of positive observations and $N = \{o \in \Omega | G(o) = N\}$ the group of negative ones. Note that here in order to simplify the notations we use P and N to denote the sets of positive and negative observations as well as to denote the letter that corresponds to the group that is assigned to the observation (this is formally an abuse of notation). A pattern aims to identify a set of attributes that have identical values for several observations in P. Of course this pattern must not appear in any observation of N.

Definition 4. *A pattern of a pdBf (P, N) is a term t_{σ^+, σ^-} such that $|P \cap T(t_{\sigma^+, \sigma^-})| > 0$ and $|N \cap T(t_{\sigma^+, \sigma^-})| = 0$.*

Given a term t, $Var(t_{\sigma^+, \sigma^-})$ is the set of attributes (also called variables) defining the term $(Var(t_{\sigma^+, \sigma^-}) = \{x_i | i \in \sigma^+ \cup \sigma^-\})$ and $Lit(t_{\sigma^+, \sigma^-}) = \{x_i \cup \bar{x}_j | i \in \sigma^+, j \in \sigma^-\}$ the set of literals (i.e. a logic variable or its complement) in t_{σ^+, σ^-}. Given a pattern p, the set $Cov(p) = P \cap T(p)$ is said to be covered by the pattern p.

Example 1. Let us recall the introductory example.

Observations	Groups	Attributes							
		α	β	γ	δ	ϵ	τ	η	κ
1	P	1	1	0	0	1	0	0	1
2		0	1	0	1	0	1	0	1
3		1	1	0	0	1	0	0	0
4		0	0	0	1	0	0	0	1
5	N	1	0	1	0	0	1	0	0
6		0	1	0	1	1	0	1	1
7		1	0	0	1	1	1	0	1
8		1	1	1	0	0	1	1	0

$p_1 = \alpha \wedge \beta \wedge \neg \gamma$ and $p_2 = \delta \wedge \neg \epsilon$ are two patterns covering respectively observations 1 and 3 (p_1) and 2 and 4 (p_2).

Let us consider now $p_3 = \neg \delta \wedge \epsilon$. p_2 and p_3 are two patterns using identical attributes: $Var(p_2) = Var(p_3)$ but $Lit(p_2) \neq Lit(p_3)$. $p_2 \cup p_3$ cover the positive group (since $Cov(p_2) \cup Cov(p_3) = P$) with only two attributes.

We consider different types of patterns:

Definition 5. *A pattern p is called* prime *if and only if the removal of any literal from $Lit(p)$ results in a term which is not a pattern.*

Obviously, a pattern is prime if and only if the removal of any variable from $Var(p)$ results in a term which is not a pattern. In Example 1, $p_2 = \neg f \wedge \neg g$ is a prime pattern. $p_4 = \neg a \wedge b \wedge \neg c$ is not prime because pattern $p_1 = \neg a \wedge b$ is prime.

Definition 6. *A pattern p_1 is called* strong *if there does not exist a pattern p_2 such that $Cov(p_1) \subset Cov(p_2)$.*

In Example 1, $p_4 = \neg \tau \wedge \neg \eta$ is a strong pattern. $p_3 = \neg \delta \wedge \epsilon$ is not strong because p_4 is a pattern and $Cov(p_3) = \{1, 3\} \subset Cov(p_4) = \{1, 3, 4\}$.

Definition 7. *A pattern p_1 is called* strong prime *if and only if*

1. *p_1 is a strong pattern and,*
2. *if there exists a pattern p_2 such as $Cov(p_2) = Cov(p_1)$ then p_1 is prime.*

In [16], it has been proved that a pattern is strong prime if and only if it is both strong and prime. Therefore, the set of all strong prime patterns is the intersection of the set of all prime patterns and the set of all strong patterns.

In the experiments, we are interested in covering the whole set of observations P by a subset of prime and strong prime patterns. Such a complete cover of group P is considered as a solution for the pattern approach. This cover is thus considered as the justification of group P from the pattern point of view.

3   Computation of Prime Patterns and Group Covers

3.1   Boros's Algorithm

In [7], Boros et al. presented an algorithm to generate the prime patterns with size smaller or equal to a given degree D. D has to be set to the highest degree D_{max} when all prime patterns are requested. This degree is not known unless generating the set of all prime patterns. Therefore, setting $D = n$, i.e. the number of variables, insures that $D \geqslant D_{max}$. The algorithm returns thus the set of all prime patterns.

Note that the algorithm generates patterns in ascending order. At each step, the patterns of size E are computed as well as a set of terms C_E to determine the patterns of size $E + 1$. The set C_E is the set of terms of size E that cover both positive and negative observations. Since a term does not cover positive observation, the term cannot be turned into a pattern by adding a literal. Hence a pattern of size greater than E is necessarily a term belonging to the set C_E to which we have added literals. The algorithm test all the combinations between the terms of C_{E-1} and the literals to determine patterns of size E, put them in a set P, and determine also the set C_E.

Algorithm 1 presents the pseudo-code of this algorithm.

Each pattern of size E determined by the algorithm comes from a term of the set C_{E-1} for which we have added a single literal. Since the set C_{E-1} contains only terms who are not patterns, the determined patterns of size E are necessarily prime patterns. Moreover, since the terms that are not contained by C_{E-1} or by P are only the terms that can never become patterns by the addition of literals, and since we test all the combinations between the terms of C_{E-1} and the literals, we necessarily determine the set of all patterns of size E. So the set P contains and contains only all prime patterns.

Note that, for an instance with x observations and n variables, the complexity is:

$$\mathcal{O}(3^n \times n \times (2^{n-\lfloor \frac{n}{3} \rfloor} \times \frac{n!}{\lfloor \frac{n}{3} \rfloor!(n-\lfloor \frac{n}{3} \rfloor)!} + x))$$

(3^n is the number of terms that can be created with n variables and $2^{n-\lfloor \frac{n}{3} \rfloor} \times \frac{n!}{\lfloor \frac{n}{3} \rfloor!(n-\lfloor \frac{n}{3} \rfloor)!}$ is the maximal number of prime patterns).

3.2 Accelerated Algorithm

In [10] we propose a new accelerated algorithm that uses the computation of all non dominated solutions of the BDCP problem in order to compute prime patterns.

In LAD, a pattern must cover at least one positive observation and no negatives ones, whatever the others positive observations. The main idea is therefore to work on observations of group P one by one. By generating the set of patterns covering each positive observation one by one, we obtain the set of all patterns. Similarly we obtain the set of all prime patterns. By generating the set of prime patterns of each positive observation,

Let us consider a given positive observation. A prime pattern can be deduced by determining the smallest set of attributes that discriminates this observation with regards to the negative observations (i.e. the non-dominated solution of the associated BDCP for which the group P contains only this positive observation).

Note that if $|P| = 1$, the set of all non-dominated solutions of the BDCP has the same attributes than the set of all prime patterns that cover the single observation of P. No attribute can be deleted.

By modifying our problem to keep only one observation in P, and by obtaining a non-dominated solution S of the BDCP (thanks to one of the previously mentioned algorithms for example), we can transform S into prime pattern: each attribute a of S appears positively (resp. negatively) in p if $D[o, a] = 1$ (resp. $D[o, a] = 0$). This transformation is insured by the Transformation_Pattern procedure in Algorithm 2.

Algorithm 1. Computation of all prime patterns using Boros's Algorithm [7]
(PPC_1).

Data: D the maximum degree of patterns that will be generated.
Result: P_D the set of prime patterns smaller than or equal to D.
$P_0 = \emptyset$
$//C_i$ is the set of terms of degree i that can become patterns, which cover both
  positive and negative observations.
$C_0 = \{\emptyset\}$
for $i = 1$ to D **do**
  $\quad P_i = P_{i-1}$
  $\quad C_i = \emptyset$
  $\quad$ **forall** $t \in C_{i-1}$ **do**
  $\quad\quad$ p=maximum index of variables in t
  $\quad\quad$ **for** $s = p + 1$ to n **do**
  $\quad\quad\quad$ **forall** $l \in \{x_s, \bar{x}_s\}$ **do**
  $\quad\quad\quad\quad$ $T = t \wedge l$
  $\quad\quad\quad\quad$ **for** $j = 1$ to $i - 1$ **do**
  $\quad\quad\quad\quad\quad$ $t' = T$ with the j-th variable removed
  $\quad\quad\quad\quad\quad$ **if** $t' \notin C_{i-1}$ **then**
  $\quad\quad\quad\quad\quad\quad$ go to $\Diamond$
  $\quad\quad\quad\quad\quad$ **end**
  $\quad\quad\quad\quad$ **end**
  $\quad\quad\quad\quad$ **if** T *covers a positive vector but no negative vector* **then**
  $\quad\quad\quad\quad\quad$ $P_i = P_i \cup \{T\}$
  $\quad\quad\quad\quad$ **end**
  $\quad\quad\quad\quad$ **if** T *covers both a positive and a negative vector* **then**
  $\quad\quad\quad\quad\quad$ $C_i = C_i \cup \{T\}$
  $\quad\quad\quad\quad$ **end**
  $\quad\quad\quad\quad$ $\Diamond$
  $\quad\quad\quad$ **end**
  $\quad\quad$ **end**
  $\quad$ **end**
end
return P_D

For each observation o, we can determine the set of non-dominated solutions
of the modified BDCP and therefore all prime patterns covering this observation.
We therefore determine that for each pattern p generated in this way, $o \in Cov(p)$.

Algorithm 2 returns the set Pat of all prime patterns, and the set Cov of
coverage of all patterns $p \in Pat$. Cov is a set of elements V_p, $\forall p \in Pat$. Each
element V_p is a set of all observations covered by p.

Note that it is not necessary to compute the set Cov to generate the set Pat.
Hence, each step that involves the set Cov can be removed.

As Boros Algorithm, it is possible to compute only prime patterns with a size
smaller than a given degree D. For this, it is sufficient to use a algorithm that
determines solutions of the BDCP that will generate only the solutions with a
size smaller than D (for example Algorithm NDS [9]).

Algorithm 2. Prime Patterns Computation (PP).

Data: D: matrix of data, with two groups {P,N}.
Result: Pat: set of all prime patterns
Result: Cov: set of covers of each prime pattern.
$Pat = \emptyset$
$Cov = \emptyset$
forall $o \in P$ **do**
    Generate the constraint matrix C_o as if o was the only one observation in P
    $Sol=$\{set of all non dominated solutions for C_o\}
    forall $s \in Sol$ **do**
        $p=$Transformation_Pattern(s,o)
        if $p \notin Pat$ **then**
            $Pat = Pat \cup \{p\}$
            //Create a new element V_p of Cov which will be a set of
            observations covered by p.
            $V_p = \{o\}$
            $Cov = Cov \cup \{V_p\}$
        end
        else
            //V_p is already in Cov; update
            $V_p = V_p \cup \{o\}$
        end
    end
end
return Pat and Cov;

Also note, if we use the algorithm NDS for computing the set Sol, for an instance with x observations and n attributes the complexity is:

$$\mathcal{O}(x^2 \times 2^{n-\lfloor \frac{n}{3} \rfloor} \times \frac{n!}{\lfloor \frac{n}{3} \rfloor!(n-\lfloor \frac{n}{3} \rfloor)!} \times \frac{n!}{\lfloor \frac{n}{2} \rfloor!(n-\lfloor \frac{n}{2} \rfloor)!})$$

($\frac{n!}{\lfloor \frac{n}{2} \rfloor!(n-\lfloor \frac{n}{2} \rfloor)!}$ is the maximal number of non-dominated solutions).

Moreover, with both sets Pat and Cov, it is possible to compute the subset of strong prime patterns by running Algorithm 3.

4   Find a Cover for a Group with the Minimal Number of Patterns

When computing all non-dominated solutions (see [9]), many solutions are generated. Hence, it is difficult to identify the most suitable solutions with regards to user-defined criteria. Another criterion could be to minimize the size of the Boolean formula induced by solution for each group in disjunctive normal form (DNF) for sake of simplicity (as mentioned in Introduction). In DNF, conjunctions of literals are connected by the logical connector $\vee$ (disjunction), where each conjunction corresponds to a pattern. The size of the Boolean DNF formula is thus the number of patterns.

Algorithm 3. Strong Prime Patterns Computation (SPP).

Data: Cov: set of coverage of each prime pattern.
Pat: set of all prime patterns
Result: SPP: set of all strong prime patterns.
$SPP = \emptyset$
forall $p \in Pat$ **do**
　if $\nexists p' \in Pat$ $s.t.$ $Cov(p) \subset Cov(p')$ **then**
　　$\mid$　$SPP = SPP \cup \{p\}$
　end
end
return SPP;

Example 2. Let us consider again Example 1

Observations	Groups	Attributes							
		α	β	γ	δ	ϵ	τ	η	κ
1	P	1	1	0	0	1	0	0	1
2		0	1	0	1	0	1	0	1
3		1	1	0	0	1	0	0	0
4		0	0	0	1	0	0	0	1
5	N	1	0	1	0	0	1	0	0
6		0	1	0	1	1	0	1	1
7		1	0	0	1	1	1	0	1
8		1	1	1	0	0	1	1	0

$S_1 = \{\delta, \epsilon\}$ is a solution of the corresponding BDCP. The Boolean formula corresponding to solution S_1 for the group P is (in DNF): $(\neg\delta \wedge \epsilon) \vee (\delta \wedge \neg\epsilon)$. The size of this Boolean formula is 2, because it contains two Boolean conjonctions (i.e. two patterns): $(\neg\delta \wedge \epsilon)$ and $(\delta \wedge \neg\epsilon)$.

When searching for a cover by means of patterns, we may again turn to a Min-Set Cover problem since we search the smaller set of patterns that cover all observations of the group P. Using the set Cov generated by Algorithm 2, we can build a constraint matrix M for this problem, where each line i represents observations of the studied group P and each column j represents a prime pattern. $M_{[i,j]} = 1$ if the pattern j covers the observation i, 0 otherwise.

Now, finding a covering for the positive group with the minimal number of patterns leads to solving the min-set cover problem represented by the following linear program:

$$min : \sum_{i=1}^{|\mathcal{A}|} y_i$$

$$s.t. :$$

$$M \cdot Y^t \geqslant \mathbb{1}^t$$

$$Y \in \{0, 1\}^{|\mathcal{A}|}, Y = [y_1, ..., y_{|\mathcal{A}|}]$$

where Y is a Boolean vector that encodes the presence/absence of each pattern in the solution. So, we can use any classic covering algorithm as the MWNV algorithm mentioned above (see Subsect. 2.3) to solve the problem and find the minimal number of patterns that fully cover the positive group.

5   Correlation Based Feature Selection (CFS)

In this section, we describe an attributes selection technique that could be relevant in our context of binary data. Among the feature selection methods, the filtering methods consist in classifying the attributes according to an appropriate selection criterion. This criterion generally depends on the relevance (i.e. correlation) of the attribute for a given cluster (i.e., group). Given a cluster, CFS aims at computing the subset of attributes that are relevant to justify this group from a classification point of view.

The **CFS** method [15] will be based on a measure μ evaluating a set of attributes $A \subseteq \mathcal{A}$ with regards to a group of observations G, taking into account the correlation between these attributes:

$$\mu(A, G) = \frac{m \times \bar{\rho}_{G,A}}{\sqrt{m + m \times (m - 1) \times \bar{\rho}_{A,A}}}$$

where $m = |A|$. Value $\bar{\rho}_{G,A}$ is the average of the correlations between the chosen attributes and a cluster/group G:

$$\bar{\rho}_{G,A} = \frac{1}{m} \sum_{i=1}^{m} \rho_{G,A_i}$$

and $\bar{\rho}_{A,A}$ is the average of the cross-correlations between the selected attributes:

$$\bar{\rho}_{A,A} = \frac{2}{m \times (m - 1)} \sum_{i=1}^{m-1} \sum_{j=i+1}^{m} \rho_{A_i,A_j}$$

where A_i is the attribute i in A. Our purpose consists thus in determining the subset A with the highest value $\mu(A, P)$ (where P is the set of positive observations). In the experiments, we use a R library[1]. Once the attributes have been selected by CFS, we will check their relevance by running a classification method and observe if these attributes allow us to rebuild the initial groups. Moreover, we will be interested in comparing the attributes selected by CFS and the two previously described approaches.

---

[1] https://www.rdocumentation.org/packages/FSelector/versions/0.21/topics/cfs.

6   Experimental Study

In the first part of this section, given a set of instance, we present the result of the comparison between the Boros's algorithm (see Subsect. 3.1) and our accelerated algorithm (see Subsect. 3.2) in terms of execution time as it was presented in [10]. In the second and third part, our purpose is to compare the different sets of attributes computed by the different methods that have been presented. Remember that these attributes aim at characterizing the groups of data. Therefore, our experimental study can be sketched by Fig. 1.

In the second part of this section, given the same set of instances, we compare:

- the number of attributes obtained by the attributes minimization approach and by the patterns minimization approach.
- the number of required patterns for covering the positive group using the attributes computed by these two approaches.

In the third part of this section, we compare sets obtained by minimization with those obtained by the variable selection method CFS. Finally, we check if the attributes obtained by the variable selection method allows us to highlight pertinent attributes for the characterization problem (i.e., minimization).

6.1   Data Instances

In order to evaluate and compare the previously described approaches, we consider different sets of observations.

- Instances ra100_phv, ra100_phy, rch8, ralsto, ra_phv, ra_phy, ra_rep1 and ra_rep2 correspond to biological identification problems. Each observation is a pathogenic bacterial strain and attributes represent genes (e.g., resistance genes or specific effectors). These bacteria are responsible of serious plant diseases and their identification is thus important.
  The main challenge for biologists is to characterize groups of bacteria using

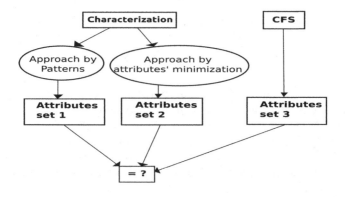

Fig. 1. Overview of experimental process.

a limited number of genes to design simple and cheap diagnosis tests [8]. The original files are available[2]. In the original data sets, several groups were considered. Here, we have considered the first group of bacteria as the positive group and the union of the other groups as the negative group. Note that similar results have been obtained when considering other groups as positive groups.

– Instance vote_r is available on the Tunedit repository[3] for machine learning tests.

The characteristics of the instances (number of observations, size of the positive group and number of attributes) are described in Table 1.

Table 1. Characteristics of the instances.

Instances	Observations	Positive group size	# Attributes
ra100_phv	101	21	50
ra100_phy	105	31	51
rch8	132	5	37
ralsto	73	27	23
ra_phv	108	22	70
ra_phy	112	31	73
ra_rep1	112	38	155
ra_rep2	112	37	73
vote_r	435	168	16

6.2 Comparision Between Algorithm of Boros and Accelerated Algorithm

The main purpose of this experiments is to compare the performance of algorithm of Boros (called algorithm PPC_1 in Subsect. 3.1) to the accelerated algorithm (called algorithm PP in Subsect. 3.2) for computing sets of prime patterns.

PP (Algorithm 2) uses the principles presented in Sect. 3.2, encoded in C++ with data structures and operators from the library *boost*[4].

PPC_1 (Algorithm 1) has been recalled in Sect. 3.1. Note that the source code of this algorithm presented in [12] was not available. It has been implemented in C++ using the same data structures and operators from the library *boost*

Experiments have been run on a computer with Intel Core i7-4910MQ CPU ($8 \times 2.90\,\mathrm{GHz}$), 31.3 GB RAM.

Table 2 provides the running times for the different algorithms whose compute the set of all prime patterns on the different instances. The first column

---

[2] http://www.info.univ-angers.fr/~gh/Idas/Ccd/ce_f.php.

[3] http://tunedit.org/repo/UCI/vote.arff.

[4] http://www.boost.org/doc/libs/1_36_0/libs/dynamic_bitset/dynamic_bitset.html.

corresponds to the name of the instance, with the number x of used attributes indicated in parenthesis (we consider different sizes for each instance to evaluate the evolution of the algorithms according to the size of the instances). The second column gives the number of prime patterns that are obtained for each instance. The following two columns give the execution times in second of our algorithm PP (Algorithm 2) and the Boros algorithm PPC_1 (Algorithm 1). Finally, the last two columns correspond to the maximum size of the patterns obtained (in number of attributes) and the execution time of PPC_1 but only looking for patterns of degree less or equal to this maximum size. In practice, we only know this size by generating all prime patterns, but Boros' algorithm can be greatly accelerated thanks to this information since it no longer has to prove that it does not exist prime pattern of size higher than this bound.

Note that the running time is limited to 24 h. " – " corresponds to execution times above this limit.

At first, we notice that the execution time of the algorithms evolves exponentially with the number of attributes. We expected this result given the complexity of the algorithms.

Then, we notice that for a low number of attributes, the two algorithms are fast, but with a significant increase of the number of attributes, only our algorithm PP can determine the set of all prime patterns in a reasonable time.

Finally, the algorithm PPC_1 with the defined maximum limit is faster than the algorithm PPC_1, and allows you to work on larger instances. However, in addition to not being able to be used in practice, it remains less efficient than our algorithm PP. The effectiveness of our algorithm PP does not come only because the algorithm PPC_1 takes a long time to prove that it has well determine all premium patterns.

6.3   Experiments on Sets of Attributes

Let us recall the main characteristics of the three methods that we consider in these experiments. Our purpose is to compare these methods in terms of selected attributes.

- MinS: computes minimal sets of attributes to identify the group P (see Sect. 2.3).
- Pattern: computes a cover of P using patterns (see Sects. 2 and 4).
- CFS: computes a set of relevant attributes for P with regards to correlation considerations (see Sect. 5).

Number of Attributes. In Table 3, we compare for (MinS and Pattern) the total number of attributes that are used. Concerning MinS, these attributes correspond to the attributes in a minimal solution. Concerning Pattern, we consider the minimal number of attributes that must be used to fully cover the set P using patterns. Note that, in both cases, several solutions may exist for the same instance.

Table 2. Results for prime patterns computation (see [10]).

Instances	Number of prime patterns	Time PP	Time PPC_1	Maximal size	PPC_1 time using max size
ra100_phv(10)	6	0.001	0.053	3	0.001
ra100_phv(20)	105	0.013	16478.822	5	11.269
ra100_phv(30)	304	0.025	–	6	40646.487
ra100_phv(40)	1227	0.089	–	8	–
ra100_phv(50)	12162	1.381	–	10	–
ra100_phy(10)	5	0.001	0.039	4	0.014
ra100_phy(20)	131	0.015	–	6	555.829
ra100_phy(30)	583	0.037	–	6	85608.555
ra100_phy(40)	1982	0.123	–	8	–
ra100_phy(51)	13112	2.119	–	10	–
rch8(15)	1	0.004	5.988	2	0.004
rch8(20)	7	0.009	4261.190	4	0.163
rch8(25)	26	0.012	–	4	0.676
rch8(30)	43	0.016	–	4	2.384
rch8(37)	131	0.021	–	6	77836.071
ralsto(10)	22	0.003	0.047	5	0.029
ralsto(15)	132	0.008	35.286	6	6.839
ralsto(20)	361	0.013	39725.665	6	436.524
ralsto(23)	1073	0.040	–	8	–
ra_phv(10)	6	0.004	0.042	3	0.004
ra_phv(25)	194	0.024	–	5	94.069
ra_phv(40)	1227	0.101	–	8	–
ra_phv(55)	28163	6.061	–	10	–
ra_phv(70)	384629	1777.386	–	14	–
ra_phy(10)	5	0.002	0.085	4	0.016
ra_phy(25)	252	0.016	–	6	8253.219
ra_phy(40)	1982	0.121	–	8	–
ra_phy(55)	23504	4.914	–	10	–
ra_phy(73)	449220	2328.949	–	14	–
ra_rep1(10)	4	0.000	0.107	4	0.013
ra_rep1(30)	11	0.000	–	4	6.385
ra_rep1(60)	415	0.021	–	7	–
ra_rep1(90)	9993	1.420	–	10	–
ra_rep1(120)	243156	1081.313	–	12	–
ra_rep1(155)	2762593	–	–	15	–
ra_rep2(10)	11	0.002	0.096	4	0.008
ra_rep2(15)	46	0.005	133.096	4	0.158
ra_rep2(20)	126	0.007	–	6	413.723
ra_rep2(25)	303	0.009	–	7	–
ra_rep2(30)	745	0.017	–	7	–
ra_rep2(35)	2309	0.060	–	9	–
ra_rep2(40)	6461	0.403	–	10	–
ra_rep2(45)	17048	2.315	–	10	–
ra_rep2(50)	43762	14.596	–	10	–
ra_rep2(55)	101026	68.378	–	11	–
ra_rep2(60)	254042	840.734	–	12	–
ra_rep2(65)	720753	4617.556	–	14	–
ra_rep2(73)	2474630	60740.333	–	15	–
vote_r(10)	169	0.060	0.876	6	0.669
vote_r(13)	1047	0.250	52.711	8	48.427
vote_r(16)	4454	0.842	4138.240	9	3466.591

And also note that the number of attributes is rather similar for both methods when considering small instances. Nevertheless, we observe a difference on ra_rep1 and ra_rep2 instances. Let us note that for ra_rep2 this difference represents only 11% of the total number of attributes of the instance, and only 6.5% for ra_rep1.

Number of Patterns for Covering *P*. In Table 4, we focus on the patterns. The value # Patterns MinS corresponds to the minimal number of patterns that are required to cover the positive group when using only the minimal set of attributes computed by MinS (in this case we recompute the patterns for the set of attributes selected by MinS, see relationships in Sect. 4). The value # Patterns is the minimal number of patterns that are necessary to cover the positive group but here the cover is based on all possible patterns.

On Table 4, we observe that the attributes computed by MinS are not suitable for finding a good set of covering patterns (based on these attributes the number of patterns for building a cover of the positive set increases). It means that the

Table 3. Number of attributes used by MinS and Pattern.

Instances	# Attributes MinS	# Attributes Pattern
ra100_phv	2	2
ra100_phy	3	4
rch8	3	3
ralsto	5	5
ra_phv	2	2
ra_phy	3	4
ra_rep1	12	22
ra_rep2	11	19
vote_r	10	14

Table 4. Number of patterns for MinS and Pattern.

Instances	# Patterns MinS	# Patterns Pattern
ra100_phv	1	1
ra100_phy	5	2
rch8	1	1
ralsto	7	3
ra_phv	1	1
ra_phy	5	2
ra_rep1	27	7
ra_rep2	25	6
vote_r	71	10

explanation provided by MinS differs from the pattern approach in terms of patterns (i.e., the attributes are different).

Comparing MinS with CFS. In Table 5, we present the following observations:

- The number of minimal solutions computed by MinS,
- The number of attributes of these minimal solutions,
- The number of attributes of the best solution computed by CFS,
- The maximum number of common attributes between the subset computed by CFS and minimal solutions of MinS. Since many solutions can maximize the number of common attributes, we also indicate in parentheses how many solutions satisfy this criterion.

Remember that the CFS method computes a subset of relevant attributes. In order to evaluate the relevance of this subset of attributes, we propose to check that they may be useful for data clustering (i.e., grouping together similar data). Therefore we use a clustering algorithm and check that the resulting clusters correspond to the initial positive and negative groups. We use here a k-means algorithm [18,21] since it is a simple, efficient and well-known clustering technique (of course the number of clusters is set to 2).

Table 5. MinS vs. CFS.

Instances	# MinS sol	# att. in min sol.	# att. CFS (accur.)	# com. att. (# sol)
ra100_phv	1	2	4 (0.465)	0 (1)
ra100_phy	1	3	1 (0.562)	1 (1)
rch8	1	3	6 (0.439)	0 (1)
ralsto	5	5	3 (0.288)	2 (1)
ra_phv	1	2	3 (0.519)	0 (1)
ra_phy	1	3	1 (0.589)	1 (1)
ra_rep1	134	12	6 (0.393)	3 (1)
ra_rep2	106	11	11(0.536)	3 (26)
vote_r	1	10	4 (0.706)	3 (1)

The accuracy of the clustering is evaluated according to the similarity between predicted cluster c and the real original group r. Given n observations we consider two n dimensional vectors C and R such as c_i is the predict group of observation i and r_i corresponds to its real original group. We define the accuracy as:

$$Acc(C, R) = |2 \times (\frac{1}{n} \sum_{i=1}^{n} (r_i - c_i)^2) - 0.5|$$

Note that $Acc(C, R) \in [0, 1]$. Values close to 1 correspond thus to a high accuracy.

For each instance the accuracy of the clustering based on the attributes selected by *CFS* is presented between parentheses. Since k-means is a statistical method we repeat the clustering process and evaluate the mean value of accuracy over 20 independent runs. The last column corresponds to the maximal number of common attributes between the subset computed by CFS and minimal solutions computed by *MinS*. Since several solutions may maximize the number of common attributes we also indicate in parentheses how many solutions satisfy this criterion.

Using an attribute selection technique such as CFS in order to explain our data by clustering techniques (i.e., distance based methods) appears not really relevant here. Note that attributes selected by *CFS* are different from attributes computed for *MinS*. Moreover, *MinS* reduces the number of attributes used for characterization. Note that we have also performed clustering using the attributes computed in *MinS* solutions leading also to poor clustering accuracy (but *MinS* is definitely not a feature selection method since it searches for combinations of attributes to characterize data).

Better results are obtained for the vote_r instance, where 3 of the 4 attributes selected by *CFS* are also involved in the unique minimal solution for *MinS*. Moreover, the accuracy is higher (0.7). Nevertheless *CFS* selects only 4 attributes, while 10 are necessary to characterize the instance in the *MinS* approach. Therefore this feature selection technique could not really be used to improve the characterization problem's results.

Evaluation of the Attributes for *MinS* and *Pattern*. In Tables 6 and 7, we evaluate the relevance of attributes using scores. We want to assess the ability of an attribute to discriminate observations according to the initial groups. Given an attribute a, the score denoted $score(a)$ is computed by counting the number of observations where the value of this attribute differs in a same group.

$$score(a) = |2 \times (\frac{1}{n} \sum_{j=1}^{n} (g_j - a_j)^2) - 0.5|$$

where:

- a_j is the value of attribute a for observation j,
- $g_j \in \{0, 1\}$ is the group assigned to observation j.

Note that an attribute that fully discriminates the groups (i.e., whose values will be always identical or always different for all observations of the two groups) has a score value of 1. Given a set of attributes (for instance representing a *MinS* solution or a *Pattern* solution), its score is the average of the scores of its attributes.

In Table 6, we focus on *MinS*:

- The first column is the best possible score obtained by a solution among the set of minimal solutions.

- In the second column, in order to evaluate this best score, we compute the ratio of it with regards to the maximal score that can be obtained when selecting the same number of attributes but using the best attributes with regards to the score function. A ratio of 1 means that the minimal solution is built using only attributes with highest scores.
- The third column indicates the best possible score among attributes that do not appear in any solution (i.e., which are not selected by MinS).
- The fourth column indicates the best possible score among the whole set of attributes.

Table 6. Attributes Scores for MinS.

Instances	Score sol max	Ratio	Max att no sol	Max att
ra100_phv	0.624	0.663	0.96	0.96
ra100_phy	0.549	0.653	0.829	0.905
rch8	0.323	0.35	0.924	0.924
ralsto	0.43	0.602	0.836	0.89
ra_phv	0.648	0.686	0.963	0.963
ra_phy	0.577	0.634	0.929	0.929
ra_rep1	0.217	0.593	0.321	0.518
ra_rep2	0.239	0.636	0.357	0.5
vote_r	0.439	0.7	0.683	0.894

Except on the vote_r instance, we note that the value of the ratio is low, which shows that the scores of the MinS solutions are rather low. These solutions are therefore composed of attributes with low scores. Moreover, the best score of the attributes not appearing in any solution is close to the best score. It means that attributes with a high score do not participate to solutions. This suggests that MinS solutions are not built with attributes that are the most correlated to the groups. And again, this scoring system does not allow preprocessing of data and offers information that is different from MinS.

In Table 7, we study solutions using Pattern. Since the number of attributes is variable, we indicate in the first column the number of attributes in the solution with the best scores between parentheses.

Similarly to MinS, we observe that the scores of the variables are not correlated to the requirements to build a cover by means of patterns.

Experimental Highlights

- As expected, MinS computes the smallest subsets of attributes, which is relevant when attributes are costly to generate (e.g., biological complex routines);

Table 7. Attributes Scores for `Pattern`.

Instances	Score sol (#att)	Ratio	Max att no sol	Max att
ra100_phv	0.624(2)	0.663	0.96	0.96
ra100_phy	0.467(4)	0.58	0.829	0.905
rch8	0.515(4)	0.557	0.924	0.924
ralsto	0.333(6)	0.514	0.836	0.89
ra_phv	0.648(2)	0.686	0.963	0.963
ra_phy	0.527(4)	0.581	0.893	0.929
ra_rep1	0.211(26)	0.616	0.339	0.518
ra_rep2	0.158(19)	0.447	0.5	0.5
vote_r	0.468(15)	0.930	0.531	0.894

- The proposed `Pattern` algorithm computes the smallest sets of patterns for covering the positive set P, which is useful to observe common characteristics shared by observations; attributes involved in these patterns differ from those selected by `MinS`;
- `MinS` and `Pattern` constitute indeed complementary methods for practitioners who need to better understand and analyze groups of binary data by focusing on different characterizations of the observations;
- Feature selection processes like `CFS` do not provide relevant information with regards to logical characterization or patterns computation.

7   Conclusion

In this paper, we focus on attributes in the general context of logical characterization and analysis of binary data. We have defined new algorithms to generate solution of the BDCP with minimal number of patterns. Our experiments show that patterns computation and computation of minimal solutions for the characterization problem constitute interesting and complementary alternatives to classic statistical based methods for features selection.

References

1. Alexe, G., et al.: Breast cancer prognosis by combinatorial analysis of gene expression data. Breast Cancer Res. **8**(4), 1–20 (2006). https://doi.org/10.1186/bcr1512
2. Alexe, G., Alexe, S., Axelrod, D., Hammer, P.L., Weissmann, D.: Logical analysis of diffuse large B-cell lymphomas. Artif. Intell. Med. **34**(3), 235–267 (2005). https://doi.org/10.1016/j.artmed.2004.11.004
3. Alexe, G., Alexe, S., Bonates, T.O., Kogan, A.: Logical analysis of data - the vision of Peter L. Hdammer. Ann. Math. Artif. Intell. **49**(1–4), 265–312 (2007). https://doi.org/10.1007/s10472-007-9065-2

4. Bennane, A., Yacout, S.: LAD-CBM; new data processing tool for diagnosis and prognosis in condition-based maintenance. J. Intell. Manuf. **23**(2), 265–275 (2012). https://doi.org/10.1007/s10845-009-0349-8

5. Boros, E., Crama, Y., Hammer, P.L., Ibaraki, T., Kogan, A., Makino, K.: Logical analysis of data: classification with justification. Ann. OR **188**(1), 33–61 (2011). https://doi.org/10.1007/s10479-011-0916-1

6. Boros, E., Hammer, P.L., Ibaraki, T., Kogan, A.: Logical analysis of numerical data. Math. Program. **79**(1–3), 163–190 (1997)

7. Boros, E., Hammer, P.L., Ibaraki, T., Kogan, A., Mayoraz, E., Muchnik, I.: An implementation of logical analysis of data. IEEE Trans. Knowl. Data Eng. **12**(2), 292–306 (2000)

8. Boureau, T., et al.: A multiplex-PCR assay for identification of the quarantine plant pathogen Xanthomonas axonopodis pv. phaseoli. J. Microbiol. Methods **92**(1), 42–50 (2013)

9. Chambon, A., Boureau, T., Lardeux, F., Saubion, F., Le Saux, M.: Characterization of multiple groups of data. In: 2015 IEEE 27th International Conference on Tools with Artificial Intelligence (ICTAI), pp. 1021–1028. IEEE (2015)

10. Chambon, A., Lardeux, F., Saubion, F., Boureau, T.: Accelerated algorithm for computation of all prime patterns in logical analysis of data, pp. 210–220, January 2019. https://doi.org/10.5220/0007389702100220

11. Chhel, F., Lardeux, F., Goëffon, A., Saubion, F.: Minimum multiple characterization of biological data using partially defined Boolean formulas. In: Proceedings of the 27th Annual ACM Symposium on Applied Computing, pp. 1399–1405. ACM (2012)

12. Chikalov, I., et al.: Logical analysis of data: theory, methodology and applications. In: Chikalov, I., et al. (eds.) Three Approaches to Data Analysis. ISRL, vol. 41, pp. 147–192. Springer, Heidelberg (2013). https://doi.org/10.1007/978-3-642-28667-4_3

13. Crama, Y., Hammer, P.L., Ibaraki, T.: Cause-effect relationships and partially defined Boolean functions. Ann. Oper. Res. **16**(1), 299–325 (1988)

14. Dupuis, C., Gamache, M., Pagé, J.F.: Logical analysis of data for estimating passenger show rates at air canada. J. Air Transp. Manag. **18**(1), 78–81 (2012)

15. Hall, M.A., Smith, L.A.: Feature subset selection: a correlation based filter approach (1997)

16. Hammer, P.L., Kogan, A., Simeone, B., Szedmák, S.: Pareto-optimal patterns in logical analysis of data. Discrete Appl. Math. **144**(1), 79–102 (2004)

17. Hammer, P.L., Bonates, T.O.: Logical analysis of data–an overview: from combinatorial optimization to medical applications. Ann. Oper. Res. **148**(1), 203–225 (2006)

18. Hartigan, J.A., Wong, M.A.: Algorithm as 136: a k-means clustering algorithm. J. R. Stat. Soc. Ser. C (Appl. Stat.) **28**(1), 100–108 (1979)

19. Karp, R.M.: Reducibility among combinatorial problems. In: Miller, R.E., Thatcher, J.W., Bohlinger, J.D. (eds.) Complexity of Computer Computations. IRSS, pp. 85–103. Springer, Boston (1972). https://doi.org/10.1007/978-1-4684-2001-2_9

20. Kholodovych, V., Smith, J.R., Knight, D., Abramson, S., Kohn, J., Welsh, W.J.: Accurate predictions of cellular response using QSPRD: a feasibility test of rational design of polymeric biomaterials. Polymer **45**(22), 7367–7379 (2004). https://doi.org/10.1016/j.polymer.2004.09.002

21. MacQueen, J., et al.: Some methods for classification and analysis of multivariate observations. In: Proceedings of the Fifth Berkeley Symposium on Mathematical Statistics and Probability, Oakland, CA, USA, vol. 1, pp. 281–297 (1967)

22. Makino, K., Hatanaka, K., Ibaraki, T.: Horn extensions of a partially defined Boolean function. SIAM J. Comput. **28**(6), 2168–2186 (1999). https://doi.org/10.1137/S0097539796297954

Annealing by Increasing Resampling

Naoya Higuchi[1]([✉]), Yasunobu Imamura[2], Takeshi Shinohara[1], Kouichi Hirata[1],
and Tetsuji Kuboyama[3]

[1] Kyushu Institute of Technology, Kawazu 680-4, Iizuka 820-8502, Japan
nac24nh@gmail.com, {shino,hirata}@ai.kyutech.ac.jp
[2] System Studio COLUN, Kokubu-machi 221-2, Kurume 839-0863, Japan
imamura.kit@gmail.com
[3] Gakushuin University, Mejiro 1-5-1, Toshima, Tokyo 171-8588, Japan
ori-icpram2019@tk.cc.gakushuin.ac.jp

Abstract. Annealing by Increasing Resampling (AIR, for short) is a
stochastic hill-climbing optimization algorithm that evaluates the objective function for resamplings with increasing size. At the beginning
stages, AIR makes state transitions like a random walk, because it uses
small resamplings for which evaluation has large error at high probability. At the ending stages, AIR behaves like a local search because it
uses large resamplings very close to the entire sample. Thus AIR works
similarly as the conventional Simulated Annealing (SA, for short). As a
rationale for AIR approximating SA, we show that both AIR and SA
can be regarded as a hill-climbing algorithm according to objective function evaluation with stochastic fluctuations. The fluctuation in AIR is
explained by the probit, while in SA by the logit. We show experimentally that the logit can be replaced with the probit in MCMC, which is
a basis of SA. We also show experimental comparison of SA and AIR for
two optimization problems, sparse pivot selection for dimension reduction, and annealing-based clustering. Strictly speaking, AIR must use
resampling independently performed at each transition trial. However, it
has been demonstrated by experiments that *reuse* of resampling within
a certain number of times can speed up optimization without losing the
quality of optimization. In particular, the larger the samples used for
evaluation, the more remarkable the superiority of AIR is in terms of
speed with respect to SA.

Keywords: Annealing by Increasing Resampling · Simulated
Annealing · Logit probit · Metaheuristics · Optimization · Dimension
reduction · Pivot selection

1 Introduction

In multimedia data retrieval, similarity search by approximate matching is more
important than exact matching. This is because, in general, multimedia data has
a high dimensionality and is enormously affected by the so-called "the curse of

© Springer Nature Switzerland AG 2020
M. De Marsico et al. (Eds.): ICPRAM 2019, LNCS 11996, pp. 71–92, 2020.
https://doi.org/10.1007/978-3-030-40014-9_4

dimensionality". Dimension reduction such as FASTMAP [8], SIMPLE-MAP [23] and sketch [7,11,19,20] is known to provide efficient indexing and fast search for high dimensional data. It requires selecting a few undistorted axes from the original space. This optimal choice may cause hard combinatorial optimization problems.

Simulated Annealing (SA, for short) [15] is a general purpose randomized algorithm for global optimization problems and known to provide a good approximation to the global optimal solution of a given objective function in a vast search space. SA tries to make the transition to the randomly selected neighborhood from a current solution. Then, it determines whether to make the transition by the parameter T (which means temperature) and the evaluation values of solutions for the objective function.

In SA, the value of T is initially large and gradually decreases. At the beginning stages, SA behaves like the random walk with making bold transitions almost regardless of the evaluation values. At the ending stages, the value of T achieves near to zero, and then, SA determines the transitions just by the evaluation values, that is, it converges to the operation of the so-called local search.

In order to solve optimization problems for dimension reduction SIMPLE-MAP and sketch, we introduced an algorithm named *Annealing by Increasing Resampling* (*AIR*, for short) [12]. AIR is also a hill-climbing algorithm using evaluation of solutions for resampling with increasing size. It is applicable to optimization problems that use the evaluation of the objective function for samples instead of the true evaluation with high computational cost. AIR adopts the average of the evaluations for individual data in sample as the value of the objective function. Other than our motivation problems on dimension reduction, the pivot selection problem [4] also uses the average based evaluation; the solution is selected to maximize the average distance between every pair of data in the projected space for a sample.

AIR uses a resampling of the sample to evaluate a current solution and its neighbor solution. In each stage, AIR replaces the resampling with an independent one. At the beginning stages, AIR uses a small resamplings and increases gradually the size as the stage progresses. With a high probability, the evaluation for a small resampling has a large error with respect to the true evaluation for the entire sample. Therefore, the local optimality for small resamplings is not stable and then the solution moves significantly. On the contrast, at the ending stages, AIR uses large resamplings, for which the evaluation becomes very close to the true value and the local optimal is stable. Thus, AIR behaves like SA.

The advantage of AIR over SA is mainly in terms of computational cost. The difference is particularly noticeable for evaluating the objective function when the sample size is very large. This is because AIR uses small resamplings in early stages, which implies a very fast evaluation. In Addition, we note the difference between AIR and the annealing by subsampling [16,21]. The former collects data by *resampling*, that is, by sampling with forgetting the previous data, whereas the latter does by *subsampling* (without forgetting).

In the previous work [12], we introduced AIR to solve optimization problems in dimension reduction. Then, in ICPRAM 2019 [13], we gave a unified view for annealing algorithms and showed that the AIR has a wide range of application. Two algorithms of SA and AIR are generalized as a hill-climbing algorithm using evaluation of objective function with stochastic fluctuation. The fluctuation in AIR can be captured by *a normal distribution with the standard error*, which is explained by the *probit*. SA using acceptance probability by Hastings [10] has the fluctuation by the *logit*. Since the logit can be approximated by the probit as well known, we showed the theoretical unified view where AIR is an approximation of SA. Furthermore, we gave comparative experimental results for SA and AIR in two optimization problems. One is the sparse pivot selection for dimension reduction SIMPLE-MAP, which is one of the most important our motivations. Another is the annealing-based clustering problem [17], which is selected to show the applicability of AIR to other generally known optimization. From these experimental results, we observed that AIR approximates SA with smaller computational cost.

In the conference version [13] of this paper in ICPRAM 2019, all the computations in experiments of sparse pivot selection for SIMPLE-MAP (the computation of the actual distances before dimension reduction, the projection by dimension reduction, the computation of the projected distances after dimension reduction) to evaluate the objective function were performed well in every time. In this paper, we report the revised experimental results by using improved implementations of SA and AIR for speeding up so that just a small number of partial computations by maintaining reusable intermediate results of computations implies the efficient computation of the objective function. In addition, we propose AIR with REUSE to speed up AIR, where REUSE is a parameter specifying the number of times to reuse the same resampling. The definition of objective function for SIMPLE-MAP is slightly changed because the objective function used in [12,13] does not satisfy AIR's preconditions in a strict sense.

We give additional experiment results for sparse pivot selection to show the relation between SA and AIR in more detail. Then, we observe that cooling schedule of SA is compatible with increasing schedule of resampling size in AIR, by using the correspondence derived from the unified view. Hence, we can conclude that AIR is superior in computing time to SA especially for samples of larger size, which also follows from the formula derived from the unified view.

2   Annealing by Increasing Resampling as an Approximation of Simulated Annealing

This section provides the rationale that annealing by increasing resampling (AIR) [12] is an alternative algorithm for simulated annealing (SA) [15]. We show that both of SA and AIR are generalized in a hill-climbing algorithm based on objective function evaluation with stochastic fluctuation as a unified view.

Let U and S be a solution space and a sample, respectively. We assume that the *objective function* $E : U \times S \to \mathbb{R}$ of optimization problem is defined as the average evaluation of a solution x with respect to every $y \in S$. The notations are shown in Table 1. The optimization problem we consider is to get a solution $x^* \in U$ minimizing E, that is, $E(x^*, S) \le E(x, S)$ for any $x \in U$.

Table 1. Notations.

Notation	Description
S	A sample
$S' \subseteq S$	A resampling
U	A solution space
$x, x' \in U$	states
$Nb(x) \subseteq U$	The neighborhood of $x \in U$
$E : U \times S \to \mathbb{R}$	An objective function
$E(x, S')$	The evaluation value for $x \in U$ and $S' \subseteq S$
$E(x)$	The value $E(x, S)$ for the entire sample S
$t \in \mathbb{N}$	A time step $(0, 1, 2, \ldots)$
$T_t \ge 0$	A temperature at t (monotonically decreasing)
$s(t) \in \mathbb{N}$	A resampling size at t (monotonically increasing, $s(t) \le |S|$)

2.1 Acceptance Criterion in Simulated Annealing

In SA, an *acceptance probability (function)* [2] or *acceptance criterion* [22] means the condition that a transition to a state worse than the current one is permitted. In Algorithm 1, SA is given as a general procedure using the acceptance probability $P(T, \Delta E)$ for a temperature T and an evaluation difference $\Delta E = E(x') - E(x)$ between two states $x, x' \in U$.

procedure SA
　　// rand$(0, 1)$: uniform random number in $[0, 1)$
　　// P: acceptance probability
　　$x \leftarrow$ any state in U;
　　for $t = 0$ **to** ∞ **do**
　　　　$x' \leftarrow$ randomly selected state from $Nb(x)$;
　　　　$\Delta E \leftarrow E(x') - E(x)$;
　　　　$\omega \leftarrow$ rand$(0, 1)$;
　　　　if $\omega \le P(T_t, \Delta E)$ **then** $x \leftarrow x'$

Algorithm 1: Simulated Annealing.

The most commonly used acceptance probability, which is originally adopted in [15], is the following *Metropolis function* P_M [18].

$$P_M(T, \Delta E) = \min\{1, \exp(-\Delta E/T)\}.$$

In this paper, we consider a bit miner choice of acceptance probability given by *Barker function* [3] (or *heat bath function* [2]) P_B, which is introduced in the context of Boltzmann machine [1] as a special case of *Hastings function* [10].

$$P_B(T, \Delta E) = \frac{1}{1 + \exp(\Delta E / T)}.$$

When a state $x' \in Nb(x)$ is selected after the current state x, the following condition holds.

$$\omega \leq \frac{1}{1 + \exp(\Delta E / T)}.$$

From the above condition and the definition of the logit function

$$\text{logit}(\omega) = \log\left(\frac{\omega}{1 - \omega}\right),$$

we can derive the acceptance criterion given by the following inequality,

$$\Delta E + T_t \cdot \text{logit}(\omega) \leq 0. \tag{1}$$

In the minimization problem, if the neighbor state x' is "better" than the current state x, then the evaluation difference ΔE is less than zero. Note that we can regard the left-hand side of the formula (1) as ΔE with stochastic fluctuation multiplied by temperature T_t.

2.2   Acceptance Criterion in Annealing by Increasing Resampling

AIR uses a randomly selected subset $S' \subseteq S$ called a *resampling* to evaluate the objective function as the average of evaluation values for individual data in S'. If the size of S' is small, the evaluation may have a large error with respect to the true evaluation using S. On the other hand, if the size of S' is almost the same as S, the evaluation has very small error. AIR utilizes this nature of resampling-based evaluation. The outline of AIR is shown in Algorithm 2.

procedure AIR
    $x \leftarrow$ any state in U;
    for $t = 0$ **to** ∞ **do**
        $x' \leftarrow$ randomly selected state from $Nb(x)$;
        $S' \leftarrow$ randomly selected resampling $S' \subseteq S$ of S such that $|S'| = s(t)$;
        if $E(x', S') - E(x, S') \leq 0$ **then** $x \leftarrow x'$

Algorithm 2: Annealing by Increasing Resampling.

Let σ be the standard deviation of the evaluation difference $E(x, y) - E(x', y)$ for individual data $y \in S$ and suppose that $|S| = N$. Then, for a resampling

$S' \subseteq S$ with $|S'| = n$, the difference $E(x, S') - E(x', S')$ is assumed to follow the normal distribution with the mean $E(x', S) - E(x, S)$ and the standard deviation $\frac{\sigma}{\sqrt{n}} \cdot \sqrt{\frac{N-n}{N-1}}$, which is known as the standard error. The term $\sqrt{\frac{N-n}{N-1}}$ is the finite population correction factor of $\frac{\sigma}{\sqrt{n}}$. Therefore, we can explain the evaluation difference for S' by the true difference for S with stochastic fluctuation as follows.

$$E(x', S') - E(x, S') = E(x', S) - E(x, S) + \frac{\sigma}{\sqrt{n}} \cdot \sqrt{\frac{N-n}{N-1}} \cdot \text{probit}(\omega). \quad (2)$$

Here, ω is a uniformly random variable between 0 and 1 and $\text{probit}(\cdot)$ is the inverse of the cumulative distribution function of the standard normal distribution.

AIR selects a resampling $S' \subseteq S$. In each step, AIR should select a resampling S' independently from the previous one. Then, we can consider the similar approach such that the next subset is obtained incrementally by adding a small number of data in $S - S'$, instead of replacing S' with the next subset. Using this approach, the intermediate computational results for the current evaluation for sample S' can be reused at the next step, and therefore, much computational cost can be reduced. However, AIR does not adopt this approach, because the process should to be made independently at each step. In practice, by reusing the current subset S' at several times, we can improve the speed but sometimes the subsample must be replaced. We will discuss the reuse of resampling later in Sect. 4.4.

2.3 Generalized Annealing Algorithm as a Unified View

In this section, we compare the acceptance criteria of SA and AIR. Remember that the acceptance criterion of SA based on Hasting function is given as the formula (1):

$$\Delta E + T_t \cdot \text{logit}(\omega) \leq 0$$

On the other hand, for AIR, the acceptance criterion is derived from the Eq. (2) as follows.

$$\Delta E + \frac{\sigma}{\sqrt{n}} \cdot \sqrt{\frac{N-n}{N-1}} \cdot \text{probit}(\omega) \leq 0. \quad (3)$$

Both acceptance criteria (1) and (3) share the form where the left-hand side is the evaluation difference ΔE with a stochastic fluctuation.

The normal distribution is known to be approximated by a logistic distribution [5]. Figure 1 shows the approximation $\text{logit}(\omega) \approx \sigma_0 \cdot \text{probit}(\omega)$ when $\sigma_0 = 1.65$. Thus, the both acceptance criteria of SA and AIR can be displayed in the following generalized form.

$$\Delta E + \alpha(t) \cdot \varphi^{-1}(\omega) \leq 0.$$

Here, $\alpha(\cdot)$ is a monotonically decreasing function of time step t as T_t for SA and as $\frac{\sigma}{\sqrt{n}} \cdot \sqrt{\frac{N-n}{N-1}}$ for AIR. Also $\varphi^{-1}(\cdot)$ is the inverse function of the cumulative distribution function corresponding to the logit in SA and the probit in AIR. Note that the resampling size n used by AIR is specified by an increasing function $s(t)$. Hence, Algorithm 3 illustrates the unified view of SA and AIR.

procedure Unified_Annealing
$\quad x \leftarrow$ any state in U;
$\quad$ **for** $t = 0$ **to** ∞ **do**
$\quad\quad x' \leftarrow$ randomly selected state from $Nb(x)$;
$\quad\quad \omega \leftarrow$ rand$(0, 1)$;
$\quad\quad$ **if** $E(x') - E(x) + \alpha(t) \cdot \varphi^{-1}(\omega) \leq 0$ **then** $x \leftarrow x'$

Algorithm 3: Unified View of SA and AIR.

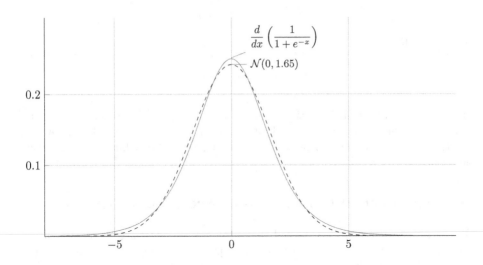

Fig. 1. Logistic distribution and normal distribution.

2.4 Compatibility of Annealing Schedules in SA and AIR

Annealing schedules are very important from the viewpoints of both efficiency and accuracy. In this section, we discuss the compatibility of the AIR's increasing schedule for resampling size with the SA's cooling schedule for temperature.

Let T_0 be the initial temperature and T_r the ratio of the current temperature T to T_0, that is:

$$T = T_0 \cdot T_r \qquad (0 < T_r < 1).$$

Also let N be the size of the entire sample S, n_0 the size of the initial resampling and n the size of the current resampling. Furthermore, let $\sigma_0 = 1.65$ and σ be the standard deviation of the evaluation difference $E(x,y) - E(x',y)$ for individual data $y \in S$.

To make the acceptance criteria (1) and (3) in SA and AIR equivalent, it is sufficient that the following equation holds, because $\mathrm{logit}(w) \approx \sigma_0 \cdot \mathrm{probit}(w)$.

$$T \cdot \sigma_0 = \frac{\sigma}{\sqrt{n}} \cdot \sqrt{\frac{N-n}{N-1}}.$$

By solving this equation for n, we have

$$n = \frac{N}{(N-1) \cdot T_0^2 \cdot T_r^2 \cdot \frac{\sigma_0^2}{\sigma^2} + 1}.$$

When $T = T_0$, it holds that $T_r = 1$ and $n = n_0$. Therefore,

$$T_0^2 \cdot \frac{\sigma_0^2}{\sigma^2} = \frac{N - n_0}{(N-1)n_0}.$$

Hence, we have:

$$s(t) = n = \frac{N}{\frac{N-n_0}{n_0} \cdot T_r^2 + 1}. \qquad (4)$$

Thus, n is given as the function of time step t, where $s(0) = n_0$. Hence, the above equation shows the compatibility between annealing schedules of SA and AIR. By using the same ratio T_r in SA and AIR, we can fairly compare two approaches.

3   Experimental Comparison of Acceptance Criteria

In the previous section, we showed that both SA and AIR are the optimization algorithms based on the evaluation of an objective function with stochastic fluctuation. Here, the fluctuation in SA is explained by using the logit, whereas that in AIR by using the probit. In this section, we introduce the several acceptance criteria and compare them. Regardless of scheduling, it is necessary to evaluate the accuracy of distribution estimation for an objective function. Therefore, instead of optimization, we use MCMC (Markov Chain Monte Calro method for Metropolis-Hastings algorithm) as the basis of SA to evaluate the accuracy.

The approximation accuracy of the estimated distribution with respect to the actual distribution of sampled points is measured by the correlation coefficient ρ. We adopt the following simple one-dimensional function as the objective function that is the weighted sum of two normal distributions and has two maxima.

$$f(x) = \frac{0.3e^{-(x-1)^2} + 0.7e^{-(x+2)^2}}{\sqrt{\pi}}.$$

Table 2. Acceptance criteria.

	Name/condition	Acceptance criteria
(i)	Metropolis acceptance criterion	$\omega \leq \min\{1, \exp(-\Delta E/T_t)\}$
(ii)	Hastings acceptance criterion	$\omega \leq \dfrac{1}{1 + \exp(\Delta E/T_t)}$
(iii)	$\varphi^{-1}(\omega) = \log(\omega)$	$\Delta E + T_t \cdot \log(\omega) \leq 0$
(iv)	$\varphi^{-1}(\omega) = \mathrm{logit}(\omega)$	$\Delta E + T_t \cdot \mathrm{logit}(\omega) \leq 0$
(v)	$\varphi^{-1}(\omega) = 1.60 \cdot \mathrm{probit}(\omega)$	$E + T_t \cdot 1.60 \cdot \mathrm{probit}(\omega) \leq 0$
(vi)	$\varphi^{-1}(\omega) = 1.65 \cdot \mathrm{probit}(\omega)$	$E + T_t \cdot 1.65 \cdot \mathrm{probit}(\omega) \leq 0$

For experiments, we also introduce the six acceptance criteria as Table 2.

As shown in Sect. 2.1, the criteria (i) and (ii) are equivalent to the criteria (iii) and (iv), respectively. On the other hand, the criteria (v) and (vi) are acceptable alternatives of the Hastings acceptance criterion when $\sigma = 1.60$ or $\sigma = 1.65$, because of the Eq. (3).

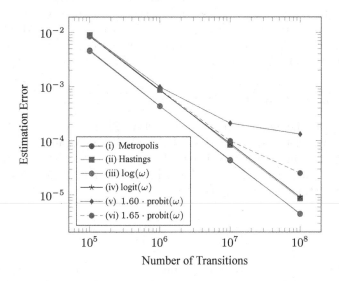

Fig. 2. Estimation errors in MCMC with 6 acceptance criteria [13].

As the burn-in phase, the first 10^5 transitions are discarded. For 1,000 discrete points of correlation coefficients, each estimation is evaluated at 10 times and we obtain the estimation error as their average. Figure 2 plots the estimation errors for each criteria.

The estimation errors of (i) Metropolis and (iii) $\log(\omega)$ are almost same, consistently with the theoretical correspondence. The same applies to (ii) Hastings and (iv) logit(ω). The criteria (v) $1.60 \cdot \text{probit}(\omega)$ and (vi) $1.65 \cdot \text{probit}(\omega)$ provide very close estimation to (ii), within 10^6 transitions. After 10^7 transitions, estimation differences become large. However, such differences cause no serious problem because 10^7 transitions are more than enough at each temperature in realistic setting of AIR. Also, it is not necessary to care about the optimal value of σ_0 because the temperature T absorbs the effect of σ, and the optimal value of σ_0 is implicitly computed in practice.

4   Experiments in Sparse Pivot Selection

In this section, we give the experimental results by applying AIR to sparse pivot selection for a dimension reduction mapping called SIMPLE-MAP [23], which is one of the most important motivation in this paper.

A dimension reduction projects data by using a mapping from a data space to a lower dimensional (projected) space. For SIMPLE-MAP, the axis of each dimension for the projected space is determined by a *pivot*, which is a point in the data space. When SIMPLE-MAP uses m pivots, m becomes the dimensionality of the projected space. In dimension reduction, it is important to reduce more dimensionalities with preserving as much distance information as possible. Here, we have to find a limited number of pivots by which SIMPLE-MAP preserves all the distances between pairs of data points as much as possible. We call this problem for SIMPLE-MAP a *sparse pivot selection*.

4.1   Dimension Reduction SIMPLE-MAP

First, we briefly introduce dimension reduction SIMPLE-MAP. Let (V, D) and (V', D') be two metric spaces such that D and D' are distance functions satisfying the triangle inequality. We denote the dimensionality of data x by $dim(x)$. We say that a mapping $\varphi : V \to V'$ is a *dimension reduction* if φ satisfies the following conditions for every $x, y \in V$.

$$dim(\varphi(x)) \leq dim(x). \tag{5}$$
$$D'(\varphi(x), \varphi(y)) \leq D(x, y). \tag{6}$$

The condition (5) reduces the dimensionality and the condition (6) makes D' the lower bound of D. A SIMPLE-MAP is based on the projection φ_p for a pivot p defined as follows.

$$\varphi_p(x) = D(p, x).$$

From the triangle inequality, the following formula holds for any $x, y \in V$.

$$|\varphi_p(x) - \varphi_p(y)| \leq D(x, y).$$

For a set $P = \{p_1, \ldots, p_m\}$ of pivots, we define a SIMPLE-MAP φ_P and a distance D' as follows.

$$\varphi_P(x) = (\varphi_{p_1}(x), \ldots, \varphi_{p_m}(x)).$$

$$D'(\varphi_P(x), \varphi_P(y)) = \max_{i=1}^{m} |\varphi_{p_i}(x) - \varphi_{p_i}(y)|.$$

Thus, φ_P is a dimension reduction when m is smaller than the dimensionality of a data space. The distance between objects projected by SIMPLE-MAP may shrink. This shrinkage, that is, the deficiency of the distance, is desired to be small for similarity search. Increasing the projected dimensionality reduces to shrinking the distance, which is strongly influenced by "the curse of dimensionality". Thus, it is important to minimize the shrinkage of the distance in a lower dimension.

Let S be a set $\{(x_1, y_1), \ldots, (x_n, y_n)\}$ of pairs of points and P a set of pivots. Then, the *distance preservation ratio (DPR)* of P for S is defined as the average of shrinkages, that is:

$$DPR(P, S) = \frac{1}{n} \sum_{i=1}^{n} \frac{D'(\varphi_P(x_i), \varphi_P(y_i))}{D(x_i, y_i)}.$$

On the other hand, in the previous works [12,13], we adopt the following old definition of DPR as the ratio of the sum of the projected distances for the sum of original distances:

$$DPR_{old}(P, S) = \frac{\sum_{i=1}^{n} D'(\varphi_P(x_i), \varphi_P(y_i))}{\sum_{i=1}^{n} D(x_i, y_i)}.$$

Whereas AIR assumes that the objective function is the average of evaluation values for individual data as discussed in Sect. 2.2, the old DPR does not meet such requirement. Therefore, we adopt the new DPR as the objective function to be maximized.

From experiments with AIR in the previous works, the assumption for the objective function seems not to be necessary. In fact, even if adopting old DPR, AIR can find the acceptable optimization results. Nevertheless, in this section, we adopt the new DPR to confirm the compatibility of optimization processes by SA and AIR.

4.2   Setting of Experiments

In this section, we give experiments for a database consisting of about 6.8 million image features. They are 2D frequency spectrums in log scale extracted from frames of 1,700 videos and the dimensionality is 64. The distance between image features is measured by an L_1 distance function. Also we fix the number m of pivots for SIMPLE-MAP to 8, which gives relatively good performance of similarity search using R-tree [9] constructed by clustering of the projected images sorted by Hilbert sort [14].

Then, we compare SA with AIR by using samples consisting of randomly selected pairs from the database. Here, the sample size N varies from 5,000 to 160,000 and the number of state transition trials (*Trials*) is from 10,000 to 640,000. The computer environment is: Intel®Core™ i7-4770K CPU 3.50 Hz, and 16 GB RAM, running Ubuntu (Windows Subsystem for Linux) on Windows 10.

4.3   Speed up by Improved Evaluation of Objective Function

We explain the improvement of the implementations of SA and AIR to the previous works. The difference between the neighbor of the state and the current state is very small as usual. In our sparse pivot selection, for the current set $P = \{p_1, \ldots, p_m\}$ of pivots, we randomly select a set P' of pivots such that the difference of P' from P is just one dimension for one pivot. That is, the difference of P' from P is one of $64m$ values. For a sample $S = \{(x_1, y_1), \ldots, (x_n, y_n)\}$ of pairs, in order to evaluate $DPR(P, S)$ of P for S, we use the following setting:

> For $i = 1, \ldots, n$, the original distance is $D(x_i, y_i)$, the images of x_i and y_i by φ_P are $\varphi_P(x_i) = (D(p_1, x_i), \ldots, D(p_m, x_i))$ and $\varphi_P(y_i) = (D(p_1, y_i), \ldots, D(p_m, y_i))$, and the projected distance is $D'(\varphi(x_i), \varphi(y_i))$.

In the previous experiments presented at ICPRAM 2019 [13], all the computations for the above setting were naively performed at each time. Since the original distances did not change in the optimization process, they were enough to be computed once at the beginning. Just a small part of images at the neighbor was different from that at the current state. Therefore, by storing the images, it was necessary to compute just a small part of images to evaluate the neighbor.

On the other hand, by adopting the above improvement, the computing time to execute SA for $N = 5,000$ and $Trials = 40,000$ is shortened to 5.6 s from 145 s. Thus, the improvement of computing just the difference of the current state and its neighbor can naturally accelerate SA. Without this acceleration, it is unfair to compare the computation time of SA with that of AIR.

As for AIR, the same acceleration is not directly applied, because the resampling is done independently from the current state for each state transition trial. However, even for AIR, if we store not only the images for data selected by resampling but also the images for other data in the entire sample, we can speed up the evaluation. AIR needs 20 s before this improvement, which is shortened to 2.3 s. We can observe the merit in speed of AIR over SA even if the improved evaluation of objective function is used.

4.4   Reuse of Resampling in AIR

In SA, whereas the transition judgment is performed by comparing the evaluation of the current state with that of the neighboring state, the difference between them is small. Therefore, when evaluating the neighbor state, we can *reuse* the intermediate computing results for the current state. Then, we can reduce the significant cost, because all the samples are used throughout in optimization.

On the other hand, in AIR, since the evaluation is performed while replacing partial samples, we cannot easily reuse the intermediate results. However, for the intermediate results, we can store and reuse the computations for the entire sample, and the computation to determine the average by summing the individual data evaluation values are only performed on partial samples. Since the cost of the aggregation is less than the SA, AIR has the cost advantage over SA.

In AIR, by performing resampling at intervals (specified by the parameter *REUSE*) without resampling at each time, we can reuse the intermediate results of computation. In contrast, since resampling is basically performed independently at each step of iteration, using a larger *REUSE* implies degrading the quality of optimization.

Table 3 shows the effect of reuse by resampling under the sample size $N = 5,000$ and the number of state transition trials *Trials* $= 40,000$, respectively. AIR with *REUSE* $= 5$ (3.2 s) is slower than AIR without *REUSE* (2.3 s). The reason is the overhead of introducing *REUSE*. Also, larger *REUSE* implies less quality of optimization. For example, the solutions for *REUSE* $= 50$ and *REUSE* $= 1,600$ are 56.4 and 55.8, respectively. For *Trials* $= 40,000$, when *REUSE* $= 1,600$, the independent resampling is performed only 25 times. By other experiments, the optimization quality is similarly degraded even if data is sequentially added to the original partial sample at the time of resampling.

Hence, in the following experiments, we use AIR with *REUSE* $= 50$, because it can optimize in relatively short time without losing the quality of optimization so much.

Table 3. Effect by Resampling with *REUSE*.

REUSE	–	5	10	20	50	100	200	400	800	1,600
Time (sec)	2.3	3.2	2.1	1.7	1.4	1.3	1.2	1.2	1.1	1.0
DPR		56.5	56.6	56.3	56.5	56.4	56.4	56.4	56.1	56.3	55.8

4.5 Annealing Schedules of SA and AIR

Next, we observe the compatibility of the schedules of SA and AIR under $N = 5,000$ and *Trials* $= 40,000$. Let *Progress* be the ratio of the number of state transition trials with respect to the total number of trials. As the cooling schedule of SA, we use the following linearly decreasing schedule with respect to *Progress*.

$$T = T_0 \left(1 - \frac{Progress}{Trials} \right).$$

By running experiments, the initial temperature $T_0 = 8.0$ for SA finds the pivots with the highest DPR. Then, the experiments on SA in this section adopt $T_0 = 8.0$. Also the initial resampling size $n_0 = 25$ for AIR behaves similarly to SA as shown in Fig. 3 under the sample size $N = 5,000$ and the number of state transitions *Trials* $= 40,000$. Here, the resampling size n is given by the Eq. (4) representing schedule compatibility using the corresponding SA temperature to initial temperature ratio $T_r = \frac{T}{T_0}$. Hence, $n_0 = 25$ gives almost the same behavior even for AIR with *REUSE* $= 50$. In Fig. 3, DPR by AIR is evaluated using the entire sample not using resampling. Each method results in approximately the same optimization progress.

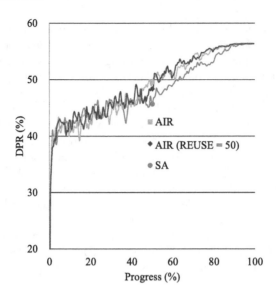

Fig. 3. Schedule compatibility of SA and AIR.

4.6 Comparison in Different Number of State Transition Trials

Table 4 represents the relationship between the number of state transition trials (*Trials*) and the optimization result. If *Trials* is 2×10^4 or more, then the optimization progresses slowly and the increase of *Trials* in any method can achieve the same level of the optimization.

Table 4. Trials and optimization result.

Trials	DPR (%)		
	SA	AIR	REUSE
1×10^4	55.23	56.03	55.88
2×10^4	56.01	56.21	56.22
4×10^4	56.35	56.53	56.44
8×10^4	56.49	56.45	56.61
16×10^4	56.58	56.60	56.68
32×10^4	56.61	56.64	56.59
64×10^4	56.67	56.74	56.73

Figure 4 illustrates the relationship between *Trials* and the computation time. As usual, we apply the local search iteration at 1,000 times for each method after annealing phase. On the other hand, in the case where the search time is minimal, the local search after the temperature reached to 0 occupies the most of the time. Therefore, the experimental results in Fig. 4 exclude the time for the local search, because it becomes an obstacle in comparing the search times.

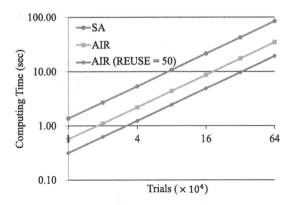

Fig. 4. Trials and computing time.

4.7 Comparison in Samples of Different Size

We compare AIR with SA varying the sample size. Note that the initial temperature T_0 of SA is independent from the sample size N. When N is increasing, it is necessary that the initial resampling size n_0 is increasing as well. On the other hand, n_0 is assumed to be fixed in practice. This is because the finite population correction factor $\sqrt{\frac{N-n}{N-1}}$ of the standard error is almost independent from N and very close to 1 when $n \ll N$.

Even a slight change of n_0 actually has a significant effect on the schedule. Figure 5 plots the optimization processes of AIR under $n_0 = 20, 25, 30, 35$ for $N = 5,000$ and $Trials = 40,000$. Then, Fig. 5 shows that only small change in n_0 affects largely optimization behaviors.

Figure 6 plots the optimization processes of AIR ($n_0 = 25$, $N = 5,000$), AIR ($n_0 = 25$, $N = 20,000$), and SA ($T_0 = 8.0$, $N = 20,000$). Then, Fig. 6 shows that the same n_0 can be used for samples of different size.

Figure 7 illustrates the temperature schedule of SA and the corresponding resampling size schedule of AIR. Here, the vertical axis is the ratio $\frac{T}{T_0}$ of temperature T to the initial temperature T_0, and the ratio $\frac{n}{N}$ of resampling size n to the entire sample size N. The evaluation cost of the objective function for sample S is usually proportional to N. The total evaluation cost in SA is proportional to

$$Trials \cdot N.$$

On the other hand, the total evaluation cost in AIR is proportional to

$$\sum_{t=0}^{Trials} n = s(t).$$

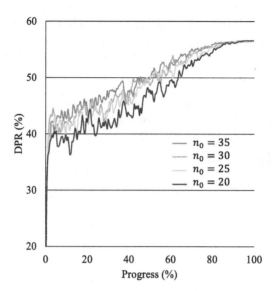

Fig. 5. Influence of initial resampling size n_0.

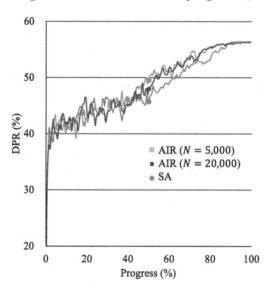

Fig. 6. Influence of sample size N.

Then, Fig. 7 shows that the difference in evaluation cost between SA and AIR becomes larger as N becomes larger.

Table 5 represents the relation between sample size and optimization quality. Then, Table 5 shows the tendency that the achieved DPR becomes smaller if the sample size is larger. This is reasonable because we fix the number of transition trials to 40,000 in these experiments.

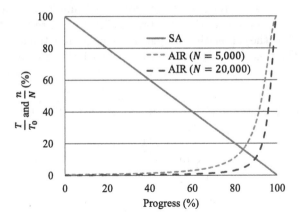

Fig. 7. Temperature and resampling size.

Table 5. Sample size and optimization result.

Sample size	DPR (%)		
	SA	AIR	REUSE
5×10^3	56.35	56.53	56.44
10×10^3	56.23	56.35	56.31
20×10^3	56.19	56.17	56.32
40×10^3	56.20	56.25	56.29
80×10^3	56.06	56.40	56.35
160×10^3	56.02	56.26	56.28

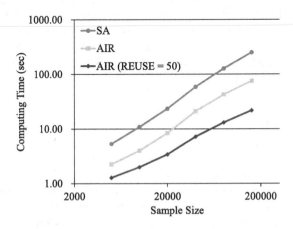

Fig. 8. Sample size and computing time.

Figure 8 illustrates the relationship between the sample size and the computation time. Table 6 represents the speed ratio of AIR with respect to SA. Then, Table 6 shows that the larger the sample size, the greater the speed advantage. In particular, the speed ratio becomes large when REUSE is used. It is about 12 for $N = 160,000$, while about 4 for $N = 5,000$.

Table 6. Speed-up ratio by AIR and AIR with REUSE to SA.

Sample size	AIR	REUSE
5×10^3	2.36	4.15
10×10^3	2.72	5.43
20×10^3	2.80	6.91
40×10^3	2.80	8.14
80×10^3	3.00	9.76
160×10^3	3.36	11.58

5   Experiments in Annealing-Based Clustering

This section focuses on a clustering problem to which we apply when comparing SA with AIR. The experimental results demonstrate the applicability of AIR to other well-known optimization problem of the sparse pivot selection. The purpose in this section is to minimize the error for sum of squares (SSE), which is commonly adopted to evaluate the objective function. Here, the clustering error for each point is the distance to the center of the cluster closest to the point.

Merendino and Celebi proposed an annealing-based clustering algorithm, named as SAGM (SA clustering based on center perturbation using Gaussian Mutation) [17]. Since they adopted two cooling schedules for SAGM, MMC (multi Markov chain) and SMC (single Markov chain), we refer SAGM using the schedule MMC and SMC to SAGM (MMC) and SAGM (SMC), respectively. As same as [17], we use 10 datasets in the UCI Machine Learning Repository [6] in the experiments. Table 7 represents the dataset description. They reported that the convergence speed of SAGM (SMC) is significantly fast and the quality of the solutions obtained by SAGM (SMC) is comparable with other SA algorithms.

We compare SAGM (MMC) and SAGM (SMC) with AIR. All the algorithms are implemented by C++. AIR uses resampling size schedules according to (4) in Sect. 2, which are corresponding to MMC and SMC. Then, we compare the quality of the solutions and the running time for the datasets. Here, the quality of the solutions is evaluated by SSE, that is, clustering with a smaller SSE means a better result. The computer environment is an Intel®Core™ i7-7820X CPU 3.60 Hz, and 64G RAM, running Ubuntu (Windows Subsystem for Linux) on Windows 10.

Table 7. Datasets (N: #points, d: #attributes, k: #classes) [13].

ID	Dataset	N	d	k
1	Ecoli	336	7	8
2	Glass	214	9	6
3	Ionosphere	351	34	2
4	Iris bezdek	150	4	3
5	Landsat	6,435	36	6
6	Letter recognition	20,000	16	26
7	Image segmentation	2,310	19	7
8	Vehicle silhouettes	846	18	4
9	Wine quality	178	13	7
10	Yeast	1,484	8	10

Tables 8 and 9 represent the quality of solutions (SSE) and the computing time obtained by SAGM and AIR using MMC and SMC schedules. All the values are the average values with parenthesized standard deviations. As for the quality of solutions, we observe no significant differences between SAGM and AIR. On the other hand, regardless of the schedule, AIR is significantly faster than SAGM for all the datasets except the 9th dataset. For the 9th dataset, which has very small number of data ($N = 178$), SAGM slightly outperforms AIR.

Table 8. Quality of solutions (sum of squared errors) [13]

Data ID	SAGM(MMC)	AIR(MMC)	Data ID	SAGM(SMC)	AIR(SMC)
1	17.55 (0.23)	**17.53** (0.20)	1	17.60 (0.29)	**17.56** (0.24)
2	**18.91** (0.69)	19.05 (0.45)	2	**18.98** (0.73)	19.08 (0.46)
3	**630.9** (19.76)	638.8 (43.42)	3	**630.9** (19.76)	646.8 (57.14)
4	6.988 (0.03)	**6.986** (0.02)	4	**6.988** (0.03)	6.991 (0.03)
5	1742 (0.01)	1742 (0.01)	5	1742 (0.01)	1742 (0.01)
6	2732 (14.17)	**2720** (4.20)	6	2738 (17.11)	**2722** (5.10)
7	411.9 (18.11)	**395.2** (10.68)	7	413.8 (19.88)	**396.2** (11.26)
8	225.7 (4.54)	**224.6** (3.82)	8	225.8 (4.65)	**224.6** (3.83)
9	37.83 (0.23)	**37.81** (0.23)	9	37.85 (0.27)	**37.82** (0.24)
10	**58.90** (1.65)	59.08 (0.74)	10	59.36 (1.61)	**59.04** (0.67)

To observe the effect of the sample size N when computing SAGM and AIR, Fig. 9 plots the ratio of the running time of SAGM for that of AIR. Then, Fig. 9 shows that the acceleration by AIR is larger for the larger samples no matter of schedules MMC and SMC.

Table 9. Average computing time (sec) [13].

Data ID	SAGM(MMC)	AIR(MMC)	Data ID	SAGM(SMC)	AIR(SMC)
1	1.727	**1.376**	1	0.282	**0.182**
2	0.719	**0.629**	2	0.125	**0.091**
3	1.549	**0.857**	3	0.281	**0.095**
4	0.139	**0.110**	4	0.022	**0.015**
5	17.40	**1.051**	5	3.216	**0.125**
6	167.4	**13.72**	6	28.82	**1.803**
7	3.285	**0.606**	7	0.523	**0.056**
8	1.298	**0.367**	8	0.219	**0.035**
9	**0.786**	0.814	9	**0.141**	0.149
10	2.854	**0.971**	10	0.497	**0.088**

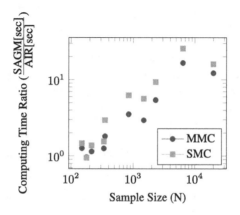

Fig. 9. Running time ratio of SAGM to AIR [13].

6 Conclusions

In this paper, we showed that AIR is an approximation of SA based on sampling from both theoretical and experimental viewpoints. The main advantage of AIR over SA is the computational cost of state transitions using resamples with a small size corresponding to state transitions at the high temperature of the conventional SA. In practice, it is necessary to increase the number of transitions at high temperatures in order to obtain stable optimization using SA. Even in such cases, AIR can produce optimization efficiently without increasing the computational cost so much. Surprisingly, the speed advantage of AIR over SA is increasing when the sample size N is increasing. This is because the initial sample size n_0 in AIR is almost unaffected by N. We expect that dimension reductions and sketches highly optimized by AIR help our important future tasks to realize of efficient similarity search for high dimensional data.

Acknowledgments. The authors would like to thank Prof. M. Emre Celebi who kindly provides us with the source codes of SAGM with both SMC and MMC schedules.

This work was partially supported by JSPS KAKENHI Grant Numbers 16H02870, 17H00762, 16H01743, 17H01788, and 18K11443.

References

1. Aarts, E., Korst, J.: Simulated Annealing and Boltzmann Machines: A Stochastic Approach to Combinatorial Optimization and Neural Computing. Wiley, Hoboken (1989)
2. Anily, S., Federgruen, A.: Simulated annealing methods with general acceptance probabilities. J. Appl. Prob. **24**, 657–667 (1987)
3. Barker, A.A.: Monte Carlo calculations of the radial distribution functions for a proton-electron plasma. Aust. J. Phys. **18**, 119–133 (1965)
4. Bustos, B., Navarro, G., Chávez, E.: Pivot selection techniques for proximity searching in metric spaces. In: Proceedings of Computer Science Society, SCCC 2001. XXI Internatinal Conference of the Chilean, pp. 33–40. IEEE (2001)
5. Demidenko, E.: Mixed Models: Theory and Applications with R, 2nd edn. Wiley, Hoboken (2013)
6. Dheeru, D., Karra Taniskidou, E.: UCI machine learning repository (2017). http://archive.ics.uci.edu/ml. University of California, Irvine, School of Information and Computer Sciences. http://archive.ics.uci.edu/ml
7. Dong, W., Charikar, M., Li, K.: Asymmetric distance estimation with sketches for similarity search in high-dimensional spaces. In: Proceedings of the 31st ACM SIGIR, pp. 123–130 (2008)
8. Faloutsos, C., Lin, K.: FastMap: a fast algorithm for indexing, data-mining and visualization of traditional and multimedia datasets. In: Proceedings of ACM SIGMOD 1995, vol. 24, pp. 163–174 (1995)
9. Guttman, A.: R-trees: a dynamic index structure for spatial searching. In: Yormark, B. (ed.) Proceedings of SIGMOD 1984, pp. 47–57. ACM Press (1984)
10. Hastings, W.K.: Monte Carlo sampling methods using Markov chains and their applications. Biometoroka **57**, 97–109 (1970)
11. Higuchi, N., Imamura, Y., Kuboyama, T., Hirata, K., Shinohara, T.: Nearest neighbor search using sketches as quantized images of dimension reduction. In: Proceedings of ICPRAM 2018, pp. 356–363 (2018)
12. Imamura, Y., Higuchi, N., Kuboyama, T., Hirata, K., Shinohara, T.: Pivot selection for dimension reduction using annealing by increasing resampling. In: Proceedings of Lernen, Wissen, Daten, Analysen (LWDA 2017), pp. 15–24 (2017)
13. Imamura, Y., Higuchi, N., Kuboyama, T., Hirata, K., Shinohara, T.: Annealing by increasing resampling in the unified view of simulated annealing. In: Proceedings of International Conference on Pattern Recognition Applications and Methods, (ICPRAM 2019), pp. 173–180 (2019)
14. Imamura, Y., Shinohara, T., Hirata, K., Kuboyama, T.: Fast hilbert sort algorithm without using hilbert indices. In: Proceedings of the Similarity Search and Applications - 9th International Conference, SISAP 2016, Tokyo, Japan, 24–26 October 2016, pp. 259–267 (2016)
15. Kirkpatrick, S., Gelatt Jr., C.D.: Optimization by simulated annealing. Science **220**, 671–680 (1983)
16. van de Meent, J.W., Paige, B., Wood, F.: Tempering by subsumpling. Technical report. arXiv:1401.7145v1 (2014)

17. Merendino, S., Celebi, M.E.: A simulated annealing clustering algorithm based on center perturbation using Gaussian mutation. In: Proceedings of FLAIRS Conference, pp. 456–461 (2013)
18. Metropolis, N., Rosenbluth, A.W., Rosenbluth, M.N., Teller, A.H.: Equation of state calculations by fast computing machines. J. Chem. Phys. **21**, 1086–1092 (1953)
19. Mic, V., Novak, D., Zezula, P.: Speeding up similarity search by sketches. In: Proceedings of SISAP 2016, pp. 250–258 (2016)
20. Müller, A., Shinohara, T.: Efficient similarity search by reducing I/O with compressed sketches. In: Proceedings of SISAP 2009, pp. 30–38 (2009)
21. Obermeyer, F., Glidden, J., Jones, E.: Scaling nonparametric Baysian inference via subsample-annealing. In: Proceedings of AISTATS 2014, pp. 696–705 (2014)
22. Schuur, P.C.: Classification of acceptance criteria for the simulated annealing algorithm. Math. Oper. Res. **22**, 266–275 (1997)
23. Shinohara, T., Ishizaka, H.: On dimension reduction mappings for approximate retrieval of multi-dimensional data. In: Arikawa, S., Shinohara, A. (eds.) Progress in Discovery Science. LNCS (LNAI), vol. 2281, pp. 224–231. Springer, Heidelberg (2002). https://doi.org/10.1007/3-540-45884-0_14

Applications

Implications of Z-Normalization in the Matrix Profile

Dieter De Paepe$^{(\boxtimes)}$ (ID), Diego Nieves Avendano$^{(\boxtimes)}$ (ID), and Sofie Van Hoecke$^{(\boxtimes)}$ (ID)

IDLab, Ghent University – imec, Zwijnaarde Technologiepark, Zwijnaarde, Belgium
{dieter.depaepe,diego.nievesavendano,sofie.vanhoecke}@ugent.be
http://idlab.ugent.be

Abstract. Companies are increasingly measuring their products and services, resulting in a rising amount of available time series data, making techniques to extract usable information needed. One state-of-the-art technique for time series is the Matrix Profile, which has been used for various applications including motif/discord discovery, visualizations and semantic segmentation. Internally, the Matrix Profile utilizes the z-normalized Euclidean distance to compare the shape of subsequences between two series. However, when comparing subsequences that are relatively flat and contain noise, the resulting distance is high despite the visual similarity of these subsequences. This property violates some of the assumptions made by Matrix Profile based techniques, resulting in worse performance when series contain flat and noisy subsequences. By studying the properties of the z-normalized Euclidean distance, we derived a method to eliminate this effect requiring only an estimate of the standard deviation of the noise. In this paper we describe various practical properties of the z-normalized Euclidean distance and show how these can be used to correct the performance of Matrix Profile related techniques. We demonstrate our techniques using anomaly detection using a Yahoo! Webscope anomaly dataset, semantic segmentation on the PAMAP2 activity dataset and for data visualization on a UCI activity dataset, all containing real-world data, and obtain overall better results after applying our technique. Our technique is a straightforward extension of the distance calculation in the Matrix Profile and will benefit any derived technique dealing with time series containing flat and noisy subsequences.

Keywords: Matrix profile · Time series · Noise · Anomaly detection · Time series segmentation

1 Introduction

With the lower cost of sensors and according rise of IoT and Industrial IoT, the amount of data available as time series is rapidly increasing due to rising interest of companies to gain new insights about their products or services, for example to do pattern discovery [15], user load prediction or anomaly detection [19].

This work has been carried out in the framework of the Z-BRE4K project, which received funding from the European Union's Horizon 2020 research and innovation programme under grant agreement no. 768869.

M. De Marsico et al. (Eds.): ICPRAM 2019, LNCS 11996, pp. 95–118, 2020.
https://doi.org/10.1007/978-3-030-40014-9_5

The Matrix profile is state-of-the-art technique for time series data that is calculated using two time series and a provided subsequence length. It is a one-dimensional series where each data point at a given index represents the Euclidean distance between the z-normalized (zero mean and unit variance) subsequence starting at that index in the first time series and the best matching (lowest distance) z-normalized subsequence in the second time series. Both inputs can be the same, meaning matches are searched for in the same time series. The Matrix Profile Index, which is calculated alongside the Matrix Profile, contains the location of the best match (in the second series) for each subsequence.

The Matrix Profile can be used to find the best matching subsequence in a series, i.e. motif discovery, or to find the subsequence with the largest distance to its nearest match, i.e. discord discovery. It also serves as a building block for other techniques such as segmentation [7], visualizing time series using Multidimensional Scaling [22] or finding gradually changing patterns in time series [24].

The usage of the z-normalized Euclidean distance can be explained by two factors. First, the MASS algorithm [14] was a known method to calculate the z-normalized distance between a sequence of length m and all subsequences obtained by sliding a window of length m over a longer sequence of length n. MASS was a vital part of the original method to calculate the Matrix Profile in reasonable time. Secondly, the z-normalized Euclidean distance can be seen as a two-step process to compare the *shape* of two sequences: the z-normalization transforms each sequence to their normal form, which captures their shape, after which the Euclidean distance compares both shapes. This makes the Matrix Profile well suited for finding patterns in data where a wandering baseline is present, as often occurs in signals coming from natural sources or due to uncalibrated sensors, or where patterns manifest with different amplitudes, which can occur by subtle changes in the underlying system or when comparing signals from different sources.

Although z-normalization is important when comparing time series [10], it has one major downside: when dealing with flat sequences, any fluctuations (such as noise) are enhanced, resulting in high values in the Matrix Profile. This behavior conflicts with our human intuition of similarity and can have an adverse effect on techniques based on the Matrix Profile. A preliminary example of this can be seen in Fig. 2, where a discord that is easily detectable using the Matrix Profile becomes hidden once noise is added to the signal. Previous literature has mainly avoided cases with series containing flat and noisy regions, most likely due to this effect.

This paper is an extended version of our previous work [4]. In this version, we diverge less on the many merits of the Matrix Profile and instead discuss several properties of the z-normalized distance relevant for the Matrix Profile. Furthermore, we have used a new dataset from Yahoo! Webscope in our anomaly detection use case and introduced a new visualization use case.

This paper is structured as follows: Sect. 2 lists literature related to the Matrix Profile. Section 3 discusses several properties of the z-normalized Euclidean distance that are either directly relevant when using the Matrix Profile or are used in our main contribution. Section 4 provides detail on the effect of flat, noisy subsequences in the Matrix Profile, as well as our solution to compensate for this effect. We demonstrate our technique for anomaly detection in Sect. 5, for semantic segmentation in Sect. 6, on data visualization in Sect. 7 and conclude our work in Sect. 8.

2   Related Work

In this section, we focus on works related to the Matrix Profile, introduced by Yeh et al. [23] as a new time-series analysis building block, together with the STAMP and STAMPI algorithm to calculate the Matrix Profile in batch or incremental steps respectively.

Internally, STAMP uses the z-normalized Euclidean distance metric to compare subsequences. Originally, all subsequences were compared using the MASS algorithm [14] allowing the Matrix Profile to be calculated in $O(n^2 \log n)$, with n being the length of the series. The later introduced STOMP and SCRIMP algorithms [26, 27] reduced the runtime to $O(n^2)$ for both batch and incremental calculation respectively by applying dynamic programming techniques.

Various variations or enhancements of the Matrix Profile have been published. When users want to track the best earlier and later match of each subsequence, rather than the best global match, the left and right Matrix Profile can be calculated instead [24]. The Multidimensional Matrix Profile tracks the best matches between time-series containing multiple channels [20]. Zhu et al. have suggested a way to calculate the Matrix Profile when the data contains missing values, using knowledge about the range of the data [25]. Lastly, we presented the Contextual Matrix Profile [5] as a generalization of the Matrix Profile that is capable of tracking multiple matches over configurable time spans.

Different distance measures have also been proposed for the Matrix Profile. The Euclidean distance or more general p-norm, might be useful in areas such as finances, engineering, physics or statistics [1]. A distance measure that performs a non-linear transformation along the time axis and can ignore the prefix or suffix of sequences being matched, based on Dynamic Time Warping, has been suggested by Silva et al. [6]. Recently, we suggested the Series Distance Matrix framework [5] as a way to easily combine different distance measures with the techniques processing these distances in a plug-and-play way.

Once the Matrix Profile and corresponding Matrix Profile index have been calculated, they can be used for motif or discord discovery. In case the user wants to focus on specific parts of the signal, for example based on time regions or high-variance periods in the signal, they can shift the Matrix Profile using the Annotation Vector [3], allowing them to find different sets of discords or motifs.

The Matrix Profile can also be used as a building block for other techniques. Time Series Chains are slowly changing patterns that occur throughout a time series and can be found by analyzing the left and right Matrix Profile [24]. Time series segmentation involves detecting changes in the underlying behavior of a time series and is possible using the offline FLUSS or online FLOSS algorithm [7], which investigate the number of arcs defined by the Matrix Profile index to detect likely transitions. Classification of time series is possible through a dictionary of identifying patterns discovered through the Matrix Profile [21]. The Matrix Profile has also been shown useful for MDS, a data exploration technique that does not work well when visualizing all subsequences in a series, by selecting representative subsequences of series [22]. Lastly, MPDist [8], a distance measure that treats sequences similar if they share many similar subsequences, is calculated using the Matrix Profile and has been used to summarize large datasets for visualisation and exploration [9].

The Matrix Profile has been used in various techniques across many domains. However, series where flat and noisy regions are present have been mostly avoided in related literature, most likely due to the issue mentioned in Sect. 1. We suspect this issue affects any Matrix Profile based technique using the z-normalized Euclidean distance, and will especially have a negative impact on techniques dealing with discord discovery (such as anomaly detection) or techniques involving matches made on flat sequences (such as the assumption of motifs being present in homogeneous regions when performing time series segmentation). To the best of our knowledge, this issue has not yet been discussed or solved prior to our work. We show how to solve this issue in Sect. 4, after first discussing several relevant properties of the z-normalized distance measure in Sect. 3.

3   Properties of the Z-Normalized Euclidean Distance

This section gathers aspects of the z-normalized Euclidean distance that are relevant for the remainder of this paper or when working with the Matrix Profile in general. Some properties listed here are obtainable through straightforward mathematical derivation of previously published properties, but have not yet been mentioned in Matrix Profile related literature, despite their high relevance.

3.1   Definition

The z-normalized Euclidean distance D_{ze} is defined as the Euclidean distance D_e between the *z-normalized* or *normal form* of two sequences, where the z-normalized form $\hat{X}$ is obtained by transforming a sequence X of length m so it has mean $\mu = 0$ and standard deviation $\sigma = 1$.

$$\hat{X} = \frac{X - \mu_X}{\sigma_X}$$

$$D_{ze}(X,Y) = D_e(\hat{X},\hat{Y}) = \sqrt{(\hat{x}_1 - \hat{y}_1)^2 + \ldots + (\hat{x}_m - \hat{y}_m)^2}$$

3.2   Link with Pearson Correlation Coefficient

The z-normalized Euclidean distance between two sequences of length m is in fact a function of the correlation between the two sequences, as originally mentioned by Rafiei [16], though without the derivation we provide below.

$$D_{ze}(X,Y) = \sqrt{2m(1 - corr(X,Y)}$$

To derive this property, we first highlight the following property of the inner product of a z-normalized sequence with itself:

$$\sigma_X^2 = \frac{\sum_i^m (x_i - \mu_X)^2}{m}$$

$$m = \sum_i^m \left(\frac{x_i - \mu_X}{\sigma_X} \right)^2$$

Using this, we can derive the equality as follows:

$$D_{ze}(X,Y)^2 = \sum_i^m \left(\frac{x_i - \mu_X}{\sigma_X} - \frac{y_i - \mu_Y}{\sigma_Y} \right)^2$$

$$= \sum_i^m \left(\frac{x - \mu_X}{\sigma_X} \right)^2 + \sum_i^m \left(\frac{y - \mu_Y}{\sigma_Y} \right)^2 - 2 \sum_i^m \left(\frac{x - \mu_X}{\sigma_X} \right) \left(\frac{y - \mu_Y}{\sigma_Y} \right)$$

$$= 2m \left(1 - \frac{1}{m} \sum_i^m \left(\frac{x - \mu_X}{\sigma_X} \right) \left(\frac{y - \mu_Y}{\sigma_Y} \right) \right)$$

$$= 2m(1 - corr(X,Y))$$

3.3 Distance Bounds

Since the correlation is limited to the range $[-1, 1]$, the D_{ze} between two sequences of length m will fall in the range $[0, 2\sqrt{m}]$, where zero indicates a perfect match and $2\sqrt{m}$ corresponds to the worst possible match.

As a result, the upper bound of $2\sqrt{m}$ can be used to *normalize distances* to the range $[0, 1]$, allowing us to compare matches of different lengths and enabling us to define and reuse thresholds to define degrees of similarity when using D_{ze}. This way, we can define a more uniform similarity threshold (e.g.: 0.3) for sequences of any length rather than specifying a threshold that is dependent on m (e.g.: a threshold of 3 for sequences of length 25, 6 for sequences of length 100 and so on). Note that Linardi et al. [12] had already pragmatically found the normalization factor $\sqrt{m}$ to compare matches of different lengths, though without making the connection to the underlying mathematics.

3.4 Best and Worst Matches

The distance bounds of Z_{ed} of 0 and $2\sqrt{m}$ correspond to correlation coefficients of 1 and -1 respectively. This means that for any sequence X of length m with $\sigma_X \neq 0$, $D_{ze}(X,Y) = 0$ and $D_{ze}(X,Z) = 2\sqrt{m}$ if:

$$Y = aX + b$$
$$Z = -aX + b$$

for any values of a and b, where $a > 0$.

3.5   Effects of Noise on Self-similarity

If we have a base sequence $S \in \mathbb{R}^m$ and two noise sequences $N \in \mathbb{R}^m$ and $N' \in \mathbb{R}^m$ sampled from a normal distribution $\mathcal{N}\left(0, \sigma_N^2\right)$, then the expected distance between the two sequences obtained by adding the noise to the base sequence can be expressed as follows:

$$X = S + N$$
$$Y = S + N'$$
$$\mathbb{E}\left[D_{ze}\left(X, Y\right)^2\right] = (2m + 2)\frac{\sigma_N^2}{\sigma_S^2 + \sigma_N^2} \tag{1}$$

Note that in (1), σ_N^2 is the variance of the noise and $\sigma_S^2 + \sigma_N^2$ is the expected variance of either noisy sequence. We apply the derivation below, originally published in our previous work [4]. For the remainder of this section, we treat the sequences as random variables.

$$\mathbb{E}\left[D_{ze}(X, Y)^2\right] = \mathbb{E}\left[(\hat{x}_1 - \hat{y}_1)^2 + \ldots + (\hat{x}_m - \hat{y}_m)^2\right]$$
$$= m \cdot \mathbb{E}\left[(\hat{x} - \hat{y})^2\right]$$
$$= m \cdot \mathbb{E}\left[\left(\frac{x - \mu_X}{\sigma_X} - \frac{y - \mu_Y}{\sigma_Y}\right)^2\right] \tag{2}$$

Since X and Y are the sum of the same two uncorrelated variables, they both have the same variance.

$$\sigma_X^2 = \sigma_Y^2 = \sigma_S^2 + \sigma_N^2 \tag{3}$$

Next, we decompose μ_X and μ_Y in the component from the original sequence μ_S and the influence of the noise. Here we use n as a random variable sampled from the noise distribution. Note that μ_S can be seen as a constant as it refers to the mean of the base sequence.

$$\mu_X = \mu_Y = \mu_S + \frac{n_1 + \ldots + n_m}{m}$$
$$= \mu_S + \mu_N \tag{4}$$
$$\mu_N \sim \mathcal{N}\left(0, \frac{\sigma_N^2}{m}\right)$$

We perform the same decomposition for x and y, where s is an unknown constant originating from the base sequence:

$$x = y = s + n$$
$$n \sim \mathcal{N}\left(0, \sigma_N^2\right) \tag{5}$$

Using (3), (4) and (5) in (2), canceling out constant terms and merging the distributions results in:

$$\mathbb{E}\left[D_{ze}(X,Y)^2\right] = m \cdot \mathbb{E}\left[\left(\frac{n_x - n_y - \mu_{N_x} + \mu_{N_y}}{\sqrt{\sigma_S^2 + \sigma_N^2}}\right)^2\right]$$

$$= m \cdot \mathbb{E}\left[(\nu)^2\right] \qquad (6)$$

$$\nu \sim \mathcal{N}\left(0, \frac{2 + 2m}{m} \cdot \frac{\sigma_N^2}{\sigma_S^2 + \sigma_N^2}\right)$$

To finish, we apply the theorem $\mathbb{E}[X^2] = \text{var}(X) + \mathbb{E}[X]^2$:

$$\mathbb{E}\left[D_{ze}(X,Y)^2\right] = (2m + 2) \cdot \frac{\sigma_N^2}{\sigma_S^2 + \sigma_N^2} \qquad (7)$$

4 Flat Subsequences in the Matrix Profile

While the utility of the z-normalized Euclidean distance as a shape-comparator has been proven by the many Matrix Profile related publications [12,22,24], results become counter-intuitive for sequences that contain subsequences that are flat, with a small amount of noise. While humans would consider such sequences as similar, the z-normalized Euclidean distance will be very high.

We can explain this effect in two ways and demonstrate this in Fig. 1, where we visualize three pairs of noisy sequences that only differ by their slope. First, considering the Euclidean distance on z-normalized sequences, we can see in Fig. 1 (middle) how the effect of noise becomes more outspoken for flatter sequences due to the normalization, resulting in a high Euclidean distance. Alternatively, we can consider the correlation of both sequences, as mentioned in Sect. 3.2. Looking at both sequences as a collection of points, shown in Fig. 1 (bottom), we can see that flatter sequences more closely resemble the random distribution of the underlying noise and are therefor less correlated. Since a correlation of zero corresponds to a z-normalized Euclidean distance of $\sqrt{2m}$, or $\frac{1}{\sqrt{2}} \approx 0.707$ if we rescale this value using the distance bounds mentioned in Sect. 3.3, we can see that uncorrelated sequences will have a high distance.

The effect of flat, noisy subsequences will have a negative effect on some use cases of the Matrix Profile. Since the flat sequences result in high Matrix Profile values where we would intuitively expect low values, we can estimate which use cases will suffer and which will not. For example, anomaly detection or discord discovery using the Matrix Profile involves finding the highest values in the Matrix Profile. When flat, noisy sequences are present, true discords may be hidden by this effect. Another example is the semantic segmentation technique using the Matrix Profile [7], this technique detects transitions in a signal by analyzing the matches of each subsequence, assuming homogeneous regions will contain many good matches. In this case, homogeneous regions containing flat and noisy sequences will result in poor matches, violating the base principle of the segmentation technique. Notably, motif detection will not suffer from this issue, assuming the user is not interested in flat motifs.

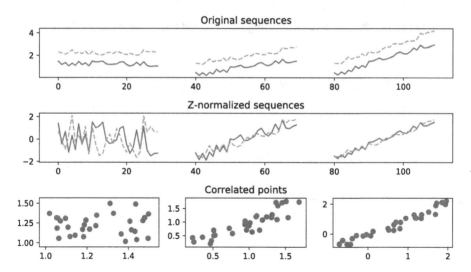

Fig. 1. Three pairs of sequences with varying slopes, each pair has the same noise profile. By looking at the effect of the noise in the z-normalized sequences, we see why the Euclidean distance will return much larger distances for flat sequences. At the bottom we see a visualization of the correlation between both sequences, where we see that the slope of the signal has a major influence on the corresponding correlation.

Next, we will discuss seemingly useful resolutions that do not actually manage to solve this effect before presenting our own solution. First however, we introduce a synthetic dataset that will serve as our running example in this section.

4.1 Running Example

We generated a sinusoid signal of 2000 samples and introduced an anomaly in one of the slopes by increasing the value of 10 consecutive values by 0.5 and create a noisy copy by adding Gaussian noise sampled from $\mathcal{N}(0, 0.01)$. The Matrix Profile for both signals was calculated using a subsequence length m of 100 and a trivial match buffer of $\frac{m}{2}$, as recommended in [23]. The signal and corresponding Matrix Profile are displayed in Fig. 2. For the noise-free signal, we see exact matches (distance equal to zero) everywhere except in the region containing the anomaly. For the noisy signal, we see how the Matrix Profile has shifted upwards, as would be expected since exact matches are no longer possible. However, we also see previously non-existing peaks in the Matrix Profile where the signal was more flat, because of this the anomaly is no longer trivial to locate automatically.

Let us briefly further investigate how the properties of the noise affect the Matrix Profile in this example. Figure 3 displays our starting sinusoidal signal with anomaly, to which Gaussian noise sampled from different distributions was added. As expected, we see that as the variation of the noise increases, the Matrix Profile becomes more deformed. The anomaly is no longer visually obvious in the Matrix Profile for noise with standard deviation of 0.05 or more. Somewhat surprising is how quickly this effect

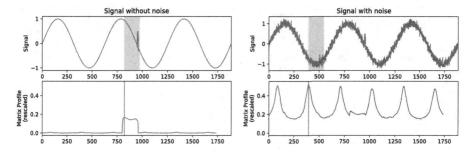

Fig. 2. Top: Sinusoid signal without (left) and with (right) added Gaussian noise. An anomaly of length 10 (red) was introduced at index 950. Bottom: Corresponding Matrix Profile for both signals, rescaled using the method of Sect. 3.3. The top discord is marked in gray. As can be seen, the presence of noise increases the Matrix Profile values of the flat regions to the degree that they now hide the true anomaly. This figure is modified from our previous work [4]. (Color figure online)

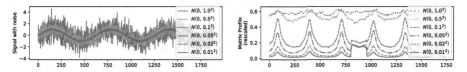

Fig. 3. A demonstration of the effect of varying degrees of noise on the Matrix Profile. Left: the sinusoidal signal with noise sampled from various Gaussian distributions. Right: The corresponding rescaled Matrix Profile. This figure is modified from our previous work [4].

becomes apparent: when the noise has a standard deviation of around 0.02 (at this point the signal-to-noise ratio is 1250 or 31 dB), the anomaly is already occasionally overtaken as the top discord by the flat subsequences (depending on the sampling of the noise).

Before coming to our solution, we will discuss why some simple, seemingly useful methods to circumvent this problem do not work.

- **Changing the Subsequence Length m:** as m becomes smaller, the effect of any anomaly on the Matrix Profile will indeed increase. However, as the subsequences become shorter and relatively flatter as a result, the effect of the noise also becomes bigger, resulting in a more eratic Matrix Profile. Increasing m will have a beneficial effect, but this is simply because the longer subsequences will become less flat in this specific example, so this is not a general solution. This approach is demonstrated in Fig. 4 (top left).
- **Ignoring Flat Sections:** ignoring subsequences whose variance is below a certain value would result in the removal of the peaks in the Matrix Profile. A first problem with this is that finding the correct cutoff value is not trivial. Secondly, this approach will not be applicable in datasets where the flat subsequences are regions of interest, either as anomalies or for finding similar subsequences, as is demonstrated in the time series segmentation of Sect. 6. This approach is visualized in Fig. 4 (middle left).

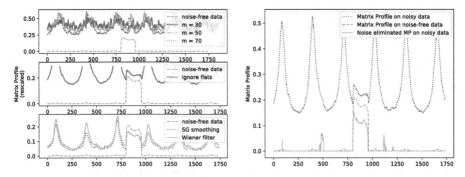

Fig. 4. Left: The effects of several seemingly useful methods to combat the effect of flat, noisy subsequences that in fact do not work. From top to bottom: reducing the subsequence length, ignoring flat sections and smoothing/filtering. None of these methods approach the noise-free Matrix Profile. Right: The effect of our noise elimination technique on the Matrix Profile. We see how the corrected Matrix Profile closely resembles the Matrix Profile of the noise-free signal. This figure is modified from our previous work [4].

- **Smoothing or Filtering:** by preprocessing the noisy signal, one could hope to remove the noise altogether. Unfortunately, unless the specifics of the noise are well known and the noise can be *completely* separated from the signal, there will always remain an amount of noise. As was shown in Fig. 3, even a small amount of noise can have a large effect on the Matrix Profile. This approach is demonstrated in Fig. 4 (bottom left).

4.2 Eliminating the Effect of Noise

Ideally, we want flat subsequences to have good matches with other flat subsequences. This would be the case if those flat subsequences were stretched and/or shifted versions of one another, as mentioned in Sect. 3.4. Unfortunately, this is not the case due to the effects of noise. We can however, still consider them to be identical, in which case we can use our derivation from Sect. 3.5, which estimates the effect of the noise on the z-normalized Euclidean distance. By subtracting this estimate during the calculation of the Matrix Profile, we are actively negating the effects of the noise. The only requirement is that we known the standard deviation of the noise that is present in the signal. This may be either known in advance or can be easily estimated by analyzing a flat part of the signal. Note that we also need the standard deviation of the subsequences being compared, but as these are already needed for the distance calculation [26,27], these are precalculated and available as part of the Matrix Profile calculation.

The algorithm is straightforward, after calculating the squared distance between a pair of subsequences using any of the existing algorithms, we subtract the squared estimate of the noise influence. We do this *before* the element-wise minimum is calculated and stored in the Matrix Profile, because this correction might influence which subsequence gets chosen as the best match. Pseudo code is listed in Algorithm 1 and can run in $O(1)$ runtime.

Algorithm 1. Algorithm for Eliminating the Effects of Noise.

Input: d: distance between subsequence X and Y
Input: m: subsequence length
Input: σ_X, σ_Y: standard deviation of subsequence X and Y
Input: σ_n: standard deviation the noise
Output: corrDist: corrected distance between subsequence X and Y

1   $corrDist = \sqrt{d^2 - (2 + m)\frac{\sigma_n^2}{max(\sigma_X, \sigma_Y)^2}}$

The only difference between this code and the formula from Sect. 3.5 is that we use the maximum standard deviation of both subsequences. When processing two fundamentally different subsequences, this choice effectively minimizes the effect of the noise elimination technique.

We demonstrate our technique on the running example in Fig. 4 (right). We see that unlike the previously methods, we can closely match the Matrix Profile of the noise-free signal. We do see some small residual spikes, which appear depending on the sampling of the noise, they are caused by local higher-than-expected noise values in that part of the signal.

After demonstrating our noise elimination technique on a limited synthetic dataset, we will use the remainder of this paper to prove the merit of our noise elimination method for several use cases using real-world datasets.

5   Use Case: Anomaly Detection

One of the original applications for the Matrix Profile is the discovery of discords, where a discord is the subsequence in a series that differs most from any other subsequence. Discords in fact correspond to the subsequences starting at the indices where the Matrix Profile is highest. When interested in the top-k discords, one can take the top-k values of the Matrix Profile where each value should be at least m index positions away from all previous discord locations. This requirement ensures we cannot select overlapping subsequences as discords, as these basically represent the same anomaly [13].

In this section we demonstrate the benefit of our noise elimination technique when performing anomaly detection utilizing real-world data from Yahoo. In our previous work [4], we performed a similar experiment using the "realAWSCloudwatch" collection from the Numenta Anomaly Benchmark [11].

We use the **Labeled Anomaly Detection Dataset of Yahoo! Webscope**, consisting of both real and synthetic time-series. For this paper we focus on the "A1Benchmark" dataset, which contains real traffic metrics from Yahoo! services, reported at hourly intervals. The benchmark consists of 67 time series with labeled anomalies, ranging from 741 to 1461 data points. The time-series vary considerable, containing diverse ranges, seasonality, trends, variance, among other properties. Figure 5 shows some examples.

Rather than classifying each point in the time series as anomalous or normal, which would involve optimizing a classification threshold, we instead score performance by

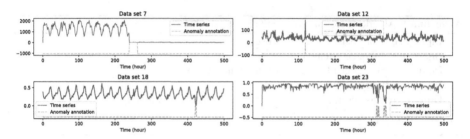

Fig. 5. Extracts of four series from the Yahoo! Webscope anomaly dataset.

counting the number of attempts needed before all anomalies in a series are reported, or until 10 incorrect guesses are made, as was done in our previous work for the Numenta benchmark [4]. This way of scoring resembles a user being alerted with suspected anomalies, measuring the capability of the algorithm to present relevant anomalies.

We perform anomaly detection by self-joining each series with a subsequence length of 24 (one day) and using the left Matrix Profile [24] for anomaly detection. The left Matrix Profile only tracks matches preceding each subsequence, similar to how streaming data is processed, and increases the chance to treat sudden changes as discords.

Since we do not know the characteristics of the noise, we would need to estimate the standard deviation using the signal. However, this is not a trivial task since we do not known in advance which signals contain noise and which do not. Making an inaccurate estimate by assuming a non-noisy signal is in fact noisy would result in poor predictions.

To detect the presence of noise, we devised a heuristic where we evaluate the Matrix Profile values of the first three days of data. If the average rescaled Matrix Profile value is above a certain threshold, we assume the start of the data is noisy and we take the median standard deviation of all subsequences in the first three days as noise parameter. Based on the first 10 datasets and leaving all other datasets as test set, we manually determined a threshold of 0.2.

This heuristic marked 34 out of 67 datasets as noisy. We compared the anomaly detection results for these 34 datasets with and without our noise elimination technique. The results are displayed in Table 1, they show that our noise elimination technique performed better for 32 out of 34 datasets. On average, the regular Matrix Profile found 36 out of a total of 84 anomalies using 291 incorrect guesses, after applying our technique this improved to 79 found anomalies using only 80 incorrect guesses. This means that on average, one in two suspected anomalies turned out to be correct!

Figure 6 shows two close-ups demonstrating the effect of the noise elimination technique. It shows the Matrix Profile having a somewhat consistent high value, whereas the noise eliminated version only increases near the actual anomalies.

Our method was unable to find all anomalies for four of the datasets. Two of these are shown in Fig. 7. In dataset 53 the first two anomalies were not detected due to their similarity with other flat series. In dataset 61 the algorithm behaves similarly to the original matrix profile due to the sudden increase in the noise level and becomes unable

Table 1. Results of anomaly detection using the Matrix Profile with and without noise elimination on the Yahoo! Webscope anomaly dataset. For each dataset, we kept guessing until all anomalies were found or 10 incorrect guesses were made. When using noise elimination we were able to find most anomalies with few attempts.

Dataset	# Anomalies	Without noise elimination		With noise elimination	
		Found anomalies	Wrong guesses	Found anomalies	Wrong guesses
1	2	0	10	2	0
2	2	1	10	2	3
4	3	0	10	3	0
5	1	0	10	1	0
6	1	0	10	1	0
8	3	0	10	3	5
10	1	0	10	1	0
11	1	0	10	1	0
12	2	1	10	2	1
14	1	1	8	1	0
17	3	1	10	3	2
19	3	0	10	3	0
21	2	1	10	2	1
22	1	1	8	1	0
23	12	5	10	12	3
24	3	3	3	2	10
25	1	1	2	1	0
31	2	0	10	2	1
32	2	2	5	2	1
33	1	0	10	1	1
40	2	1	10	2	9
41	3	1	10	3	3
42	3	0	10	3	7
43	3	2	10	3	1
45	1	0	10	1	0
48	1	0	10	0	10
50	1	1	2	1	0
53	4	4	3	2	10
58	1	0	10	1	0
61	2	0	10	1	10
62	4*	0	10	4	1
63	1	0	10	1	0
66	6	5	10	6	1
67	5	5	0	5	0
Sum	84	36	291	79	80

* Dataset 62 actually contains five anomalies, but because the first anomaly occurs within the first three days which are used to estimate the noise level, we do not consider it in the results.

to differentiate anomalies from noise. In this case the first anomaly is found but the second one is missed.

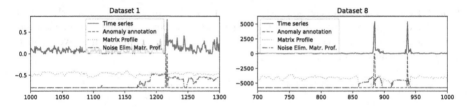

Fig. 6. Two examples displaying the beneficial effect of our noise elimination on anomaly detection. The anomalies are not noticeable in the regular Matrix Profile, but are obvious after applying noise elimination.

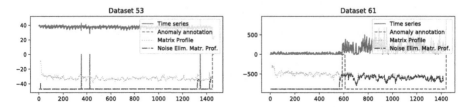

Fig. 7. Left: Dataset containing two anomalies which are not detected as they closely resemble the estimated noise. The noise eliminated Matrix Profile is zero for this segment. Right: Dataset where the original amount of noise is small but increases at one point, causing the Noise Elimination to lose its effect.

6   Use Case: Semantic Segmentation for Time Series

Semantic segmentation of time series involves splitting a time series into regions where each region displays homogeneous behavior, these regions typically correspond to a particular state in the underlying source of the signal. Applications of segmentation may include medical monitoring, computer-assisted data annotation or data analysis in general. In this section, we perform semantic segmentation on the PAMAP2 activity dataset using the Corrected Arc Curve (CAC). The CAC is calculated by the FLUSS algorithm for batch data or the FLOSS algorithm for streaming data, using the Matrix Profile index [7].

The CAC was introduced as a domain agnostic technique to perform time series segmentation on realistic datasets, with support for streaming data while requiring only a single intuitive parameter (the subsequence length to consider). During evaluation the CAC was found to perform better than most humans on dozens of datasets, allowing the authors to claim "super-human performance" [7].

The CAC is a vector of the same length as the Matrix Profile, and is constructed by analyzing the Matrix Profile Index. They consider arcs running from each subsequence to the location of its nearest match. To calculate the CAC, they compare the number of arcs running over each location against the amount of arcs expected if all match locations would be determined by uniform sampling over the entire series. This ratio is defined as the CAC, its values are strictly positive without an upper bound, but can be safely restricted to the range $[0, 1]$. Assuming homogeneous segments will display similar behavior while heterogeneous segments will not, a low CAC value is seen as evidence of a change point, though a high CAC value should not be seen as evidence of the absence of one.

We use the **PAMAP2 Activity Dataset** [17], which contains sensor measurements of 9 subjects performing a subset of 18 activities like sitting, standing, walking, and ironing. Each subject was equipped with a heart rate monitor and 3 inertial measurement units (IMU) placed on the chest, dominant wrist and dominant ankle. Each IMU measured 3D acceleration data, 3D gyroscope data and 3D magnetometer data at 100 Hz. The time series are annotated with the activity being performed by the subject and transition regions in between activities. The duration of each activity varies greatly, but most activities last between 3 to 5 min.

The PAMAP2 dataset has already been used in the context of segmentation [20], where the authors used the Matrix Profile to classify the activities in passive and active activities. At one point they note that the motif pairs in the passive actions (such as lying, sitting or standing) are less similar and therefore less useful for segmentation. The underlying problem here is the inability to detect motifs in the passive activities, because they mainly consist of flat, noisy signals. By using our method to compensate for this effect, we will be able to find the needed motifs, resulting in better segmentation results.

We applied time series segmentation on the passive activities present in the PAMAP2 dataset, focusing on the "lying", "sitting" and "standing" activities. We picked these activities since their measurements display very few patterns in the data and they are performed consecutively for all subjects, meaning we did not have to introduce time-jumps in our experiments.

We considered subjects 1 to 8 of the dataset (subject 9 had no recordings of the relevant activities) and tested both the transition from "lying" to "sitting" as well as the transition from "sitting" to "standing". For each subject, we used the 3 accelerometer signals from the IMU placed on the chest of the subject, any missing data points were filled in using linear interpolation. We calculated the CAC by self-joining each sensor channel with and without noise elimination using a subsequence length of 1000 (10 s). The standard deviation of the noise was estimated (without optimizing) by taking the 5th percentile of the standard deviations of all subsequences.

An example of the signals spanning over the 3 passive activities can be seen in Fig. 8. We emphasize it is not our goal to build the optimal segmentation tool for this specific task, but to simply evaluate the effect of our noise elimination technique on the CAC for sensor signals containing flat and noisy subsequences.

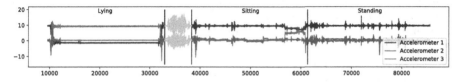

Fig. 8. Three accelerometer channels of subject 6 from the PAMAP2 dataset. We see three activities and one long transition period. No clear patterns are discernible and many flat and noisy subsequences are present. Reproduced from our previous work [4].

There is one unexpected side effect of the noise cancellation technique that needs to be corrected before calculating the CAC. Because most of the flat subsequences will have an exact match (distance equal to zero) to other flat subsequence, there will be many locations to represent the best match. However, since the Matrix Profile Index only stores one value, the selected best match will become determined by the first or last match, depending on the implementation of the Matrix Profile. This creates a pattern in the Matrix Profile Index, that actually violates the CAC's assumption of matches being spread out over a homogeneous region. Note that this effect is in fact also present in the normal Matrix Profile, but typically has goes unnoticed because multiple exact matches are extremely rare.

To prevent this effect, we need to randomly pick one of the best matches and store its location in the Matrix Profile Index. This is straightforward when using the STOMP algorithm [27], as every step in STOMP calculates all matches for one particular subsequence. If we want to calculate the Matrix Profile in an online fashion using the SCRIMP algorithm [26], where the matches for one specific subsequence are spread over many iterations, we need to utilize reservoir sampling [18] in the construction of the Matrix Profile Index. Reservoir sampling allows uniform sampling without replacement from a stream without knowing the size of the stream in advance. We use it to sample the stream of best matches. Implementing reservoir sampling requires us to store an additional vector of the same length as the Matrix Profile, to keep track of the number of exact matches that was encountered so far for each subsequence. Pseudo code to update the Matrix Profile and its indices is listed in Algorithm 2.

Algorithm 2. SCRIMP Matrix Profile Update using Reservoir Sampling.

   Input: dists: distances on diagonal calculated by SCRIMP
   Input: indices: corresponding indices of dists
   Input: numMatches: number of exact matches per subsequence
   Input: mp: part of Matrix Profile vector being updated
   Input: mpi: part of Matrix Profile Index being updated
   `/* Handle new better matches                                    */`
1 $better = dists < mp$
2 $mp[better] = dists[better]$
3 $mpi[better] = indices[better]$
4 $numMatches[better] = 1$
   `/* Handle matches equal to current best match                   */`
5 $equal = dists == mp \wedge finite(dists)$
6 $numMatches[equal] = numMatches[equal] + 1$
7 **for** i in $equal$ **do**
8   **if** $random() < 1/numMatches[i]$ **then**
9        $mpi[i] = indices[i]$

The code is straightforward. In lines 1 to 3 we update the Matrix Profile and Index if a better match was found and in line four we reset the tracked number of exact matches. In line five, we gather any matches equally good as the match being tracked in the

Matrix Profile. Line six increases the counter of any equally good matches found and line seven to nine perform the reservoir sampling to update the Matrix Profile Index for each newly found equal match.

For both experiments, the CAC was calculated using the Matrix Profile with randomly sampled indices. The activity-transition point was taken where the CAC was minimal, ignoring any values in the first and last 50 s (5 times the subsequence length), as suggested by the original paper [7]. We considered 4 segmentations per subject: one CAC for each of the three sensor channels and one obtained by averaging the individual CACs.

To evaluate the ability to predict the transition period, we define the score as the normalized distance between the predicted transition and the ground truth transition. Note that some ground truth transitions are instantaneous, while others consist of a transition period, as can be seen in Fig. 8. We also added an additional buffer period equal to the subsequence length m (10 s) before and after the transition period that we still consider as correct to consider the detection interval of the Matrix Profile. Pseudo code for our scoring function is listed in Algorithm 3, a score will range from 0 to 100, where lower is better.

Algorithm 3. Scoring Function for Semantic Segmentation.

    Input: estimate: estimated transition
    Input: trueStart, trueEnd: ground truth start and end of transition
    Input: n: length of series (both activities and transition period)
    Input: m: subsequence length / transition buffer
    Output: score
1 **if** $estimate < trueStart - m$ **then**
2 | $score = ((trueStart - m) - estimate)/n * 100$
3 **else if** $estimate > trueEnd + m$ **then**
4 | $score = (estimate - (trueEnd + b)/n * 100$
5 **else**
6 | $score = 0$

Table 2 lists the results for the segmentation when transitioning from "lying" to "sitting". For all subjects expect subject one, the results show similar of improved scores for segmentation using individual sensors as well as the combined approach when using the noise elimination technique. The average score for the individual sensors improves from 9.32 to 7.71, a modest improvement corresponding to a gain of about 8 s. The segmentation based on the combined CACs improves from 9.61 to 6.07, a gain of about 18.5 s. Note that most scores without noise elimination were already very good, leaving little room for improvement. The bad results for subject one can be explained by an incorrect early estimate which is caused by movement of the subject near the start of the "lying" activity. Note that subject one has bad scores for both techniques.

Table 3 lists the results for the transition from "sitting" to "standing". Like the previous experiment, we see similar or improved results when applying noise elimination, except for subjects one, three and eight using the first sensor series and for the combined

Table 2. Scores for the segmentation of the transition from "lying" to "sitting" using the 3 chest accelerometers from the PAMAP2 dataset for subjects 1 through 8, with and without noise elimination applied. Segmentation is performed using the CAC from a single sensor (C1, C2 and C3) and using the average of the 3 CACs (combined). Similar or better performance are achieved when applying noise elimination for all subjects except subject 1. Results are reproduced from our previous work [4].

Subject	Without noise elimination				With noise elimination			
	C1	C2	C3	Combined	C1	C2	C3	Combined
1	**5.9**	31.3	**31.9**	**31.7**	41.3	31.8	41.8	36.7
2	32.9	1.4	1.4	1.4	28.8	1.4	1.7	1.4
3	35.9	2.8	31.1	33.8	**2.4**	2.3	**2.3**	**2.3**
4	0.0	2.8	5.9	0.0	0.0	1.5	6.6	0.8
5	1.1	7.6	5.1	3.9	1.6	**1.7**	4.9	1.6
6	2.5	1.9	2.3	2.3	2.4	1.9	2.0	2.4
7	0.1	1.8	11.1	2.0	2.1	1.8	**1.9**	1.9
8	0.0	1.4	5.5	1.7	0.0	1.4	1.4	1.4
Average	9.32			9.61	7.71			6.07

Table 3. Scores for the segmentation of the transition from "sitting" to "standing" using the 3 chest accelerometers from the PAMAP2 dataset for subjects 1 through 8, with and without noise elimination applied. Segmentation is performed using the CAC from a single sensor (C1, C2 and C3) and using the average of the 3 CACs (combined). Overall, we see similar or better performance when applying noise elimination, except for the segmentation using the first channel for subject 1, 3 and 8. Results are reproduced from our previous work [4].

Subject	Without noise elimination				With noise elimination			
	C1	C2	C3	Combined	C1	C2	C3	Combined
1	**32.5**	0.0	3.6	2.2	38.7	0.0	3.7	2.2
2	36.5	37.2	36.4	37.0	**7.1**	**30.0**	32.7	**29.2**
3	**10.0**	30.2	43.1	**30.2**	43.2	**14.0**	43.7	43.2
4	7.8	1.9	1.1	1.2	**0.7**	2.0	1.3	1.3
5	13.1	0.0	28.5	10.6	13.3	1.0	**1.2**	**1.0**
6	36.1	36.6	26.9	36.6	**23.3**	**3.4**	26.5	**3.2**
7	43.1	38.0	16.5	16.5	43.4	**1.6**	**0.0**	1.6
8	**2.3**	1.0	24.8	1.0	21.1	0.0	**16.5**	1.0
Average	21.12			16.9	15.35			10.3

approach for subject three. While the overall scores are worse, the gain by enabling noise elimination is more significant. The average result for a single sensor improves from 21.12 to 15.35, corresponding to a gain of about 27 s. When using the combined approach the average score improves from 16.9 to 10.3, a gain of around 31 s.

Though limited in scope, these result indicate that the noise elimination technique improves the ability of the CAC to detect transitions in a series containing flat and noisy subsequences.

7   Use Case: Data Visualization

Data visualization is a great tool for exploring newly acquired data or for finding similar data in a larger data collection. Unfortunately, time series data for realistic use cases is typically very lengthy, making trivial visualizations useless as these will fail to simultaneously capture the overall and minute details of a series. More advanced techniques summarize the series in a way that is still useful to gain insight into the data. The visualization technique used in this section is the Contextual Matrix Profile (CMP), recently introduced by the authors [5].

The CMP is a generalization of the Matrix Profile that tracks the best match between predefined ranges whereas the Matrix Profile tracks the best match for every possible sliding window location. The distinction between the two can also be made in terms of the implicit distance matrix, defined by the pairwise distance between all subsequences of two series. Where the Matrix Profile equals the column-wise minimum of the distance matrix, the CMP consists of the minimum over rectangular areas of the distance matrix.

For this use case, we use the **Chest-Mounted Accelerometer Dataset** [2], an activity recognition dataset publicly available at the UCI repository[1]. The dataset contains data of 15 subjects performing seven different activities, measured using a chest-mounted accelerometer sampling at 52 Hz. The data is labeled with the corresponding activity, though visual inspection reveals the labels seem misaligned for some subjects. The activities performed are: Working at a computer; standing up, walking, going up/down stairs; standing; walking; going up/down stairs; walking while talking; talking while standing. We selected this dataset for this use case as it contains both activities with a periodic nature as well as passive activities where the accelerometer signal consists of mainly noise.

For the remainder of this section, we focus on the first subject of the dataset due to space constraints, though similar results were obtained for all subjects. This series comprises 52 min of data containing nine regions of activity, it is visualized in Fig. 9. In the top of the figure we see the complete dataset with annotations indicating the activity regions. At the bottom of the figure, close-ups of the signal for four different activities are displayed. While the "working at computer" consists mainly of a flat singal, a periodic pattern is visible in the channels of the other activities, with the "walking" activity having the clearest pattern. Note that the three data channels are uncalibrated, which is not an issue since the z-normalization focuses on the shape of subsequences rather than the absolute values.

The CMP for each data channel is calculated by self-joining the data, using a subsequence length of 52 (1 s) and specifying contexts of length 469 at 520 (10 s) intervals. The context length is chosen so that matches can never overlap. Calculating the CMP

---

[1] https://archive.ics.uci.edu/ml/datasets/Activity+Recognition+from+Single+Chest-Mounted+Accelerometer.

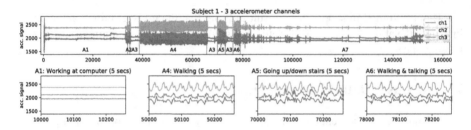

Fig. 9. Top: Uncalibrated accelerometer data (3 channels) for subject 1, sampled at 52 Hz, for a total of 52 min. The annotations A1 to A7 indicate the corresponding activity labels. The activities performed are computer work, standing up/walking/stairs, standing, walking, standing, stairs, standing, walking while talking, and talking while standing. Bottom: four extracts of 4 different activities, each 5 s long.

comes down to dividing the series in non-overlapping, contiguous 10 s windows and finding the best one second match for each pair of windows. The resulting CMP is a 312 by 312 matrix, each value representing the distance of the best match of the intervals defined by the row and column. The resulting CMPs for each data channel can be seen in Fig. 10 (left). To demonstrate the value of the visualization, we also added grayscale band to the CMP outline that corresponds to the activity labels present in the data.

To interpret a CMP visualization, one should consider both axes represent the flow of time, starting at the origin. Each value in the CMP indicates how well one region of time matched another region, low (dark) values represent good matches while high (light) values represent bad matches. Looking at Fig. 10 (left), we can observe a number of things. Most obvious is the symmetry of each CMP, this is because we performed a self-join. We also see the dark square centered at index 100 in all three channels, this indicates a period containing a repetitive pattern across all channels. This region corresponds to the "walking" activity in the dataset, as can be seen by referencing Fig. 9. Next, for channel two we see similar dark regions for the "stairs" and "walking while talking" activities, meaning there is a similarity between all three activities based on channel two. Though these activities also appear in channel one and three, they are less visually noticeable, especially the "stairs" activity in channel one could be easily missed. The same can be said for the very short "standing up/walking/stairs" activity that precedes the walking activity. One final observation is the presence of high value bands occurring across all channels to some degree between indices 0 and 60 and between 200 and 250. While the first band corresponds to the "computer work" activity, there is no clear-cut corresponding activity for the second band. Most likely, these artefacts are caused by regions with very flat signals, where the noise has a large effect on the distance calculation.

Next, we calculated the CMPs while using the noise elimination technique, keeping all other parameters equal. We estimated the noise parameter for each channel by sliding a one second window over the entire series, calculating the standard deviation for each location and taking the value corresponding to the fifth percentile, similar as we did

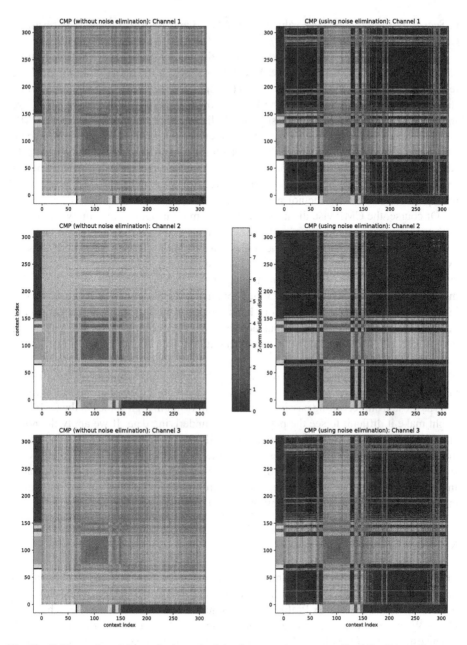

Fig. 10. CMPs produced for each channel of the dataset using z-normalized Euclidean distance without (left) and with (right) compensating for the noise. The left and bottom grayscale bar for each CMP corresponds to the different activity labels for each of the time windows. We see how the noise elimination results in large areas of exact matches for the passive activities, because of this, the different transitions between active and passive activities is clearly visible. In case the activity labels were unknown, the CMP would have given a good indication of the different regimes present in the signal.

in Sect. 6. For reference, these values were: 2.6, 2.3 and 2.7 respectively. The resulting CMPs are visualized in Fig. 9 (right).

We see the CMPs generated using noise elimination now have additional large, rectangular regions with low values. As can be seen from the activity markings, these regions correspond to the passive activities (computer work, standing and talking while standing), where the series is flat and lacks distinctive patterns. Looking in detail, we see how the activity markings almost perfectly line up with the transitions in the CMP, which is major difference with the CMPs without noise elimination. We do see some line artefacts in the final activity of the dataset for all channels, near index 190 and 280, which correspond to increases in the signal on all three channels. The questions whether or not there is an activity change occurring there is in a way debatable and may simply be a question of tweaking the noise parameter or refining the activity labels.

Of course, the CMP visualization can provide more insights than simply the difference between passive and active activities. It can also be used to differentiate between different activities, provided these activities have different underlying patterns. As an example, we can see that for the CMP of the first channel, the activities "walking" and "walking while talking" have better matches than "walking" and "stairs". This is not the case for channels two and three. When looking at the closeup data of Fig. 9, we can in fact see a more distinct pattern for channel one for the "stairs" activity. We do not quantify the ability to discern various activities as it is not in scope of this paper, our goal was simply to demonstrate the added benefit of noise elimination when visualizing data with the CMP.

To conclude, we demonstrated the effect of the noise elimination technique for data visualization using the CMP on accelerometer data for an activity dataset. Flat signals, such as those from recording passive activities, will result in high values in the CMP and might make it difficult to see the patterns of the underlying data. If we apply the noise elimination technique while calculating the CMP, the passive activities become easily discernible as regions of low values, giving the user better insight in the underlying structure of the data and allowing the user to focus more on the more salient parts. Importantly, the regions with active activities are unaffected by the noise elimination, meaning we can apply noise elimination without risk.

8 Conclusion

In this paper we explained the unintuitive behavior of the z-normalized Euclidean distance when comparing sequences that are flat and noisy, and demonstrated how this negatively affects the Matrix Profile and techniques using the Matrix Profile as building block. We discussed several properties of the z-normalized Euclidean distance, including an estimation of the effect of noise, which we use to eliminate this effect altogether.

We applied our noise elimination technique on three different use cases involving real-world data from open data sets. For anomaly detection on the Yahoo! Webscope anomaly dataset, we were able to automatically guess twice as many anomalies while utilizing less than one third of attempts when using our technique. When used for semantic time series segmentation, we showed an improved accuracy for detecting the transition between two passive activities. Finally, in our visualization use case, we showed a major change in the visualization of activity data using the Contextual Matrix Profile, allowing us to separate the underlying activities that were previously indistinguishable.

Since our technique is conceptually simple, users should be able to reason whether or not their use case will benefit from our technique. Our technique is straightforward to implement and incurs only a constant factor overhead, so it can be used by everyone using Matrix Profile related techniques working with data containing flat and noisy subsequences.

Future remains on more robust noise estimations and dealing with series where noise characteristics change over time.

References

1. Akbarinia, R., Cloez, B.: Efficient Matrix Profile Computation Using Different Distance Functions (2019). https://arxiv.org/abs/1901.05708
2. Casale, P., Pujol, O., Radeva, P.: Personalization and user verification in wearable systems using biometric walking patterns. Pers. Ubiquit. Comput. **16**, 1–18 (2012)
3. Dau, H.A., Keogh, E.: Matrix profile V: a generic technique to incorporate domain knowledge into motif discovery. In: Proceedings of the 23rd ACM SIGKDD International Conference on Knowledge Discovery and Data Mining, pp. 125–134 (2017)
4. De Paepe, D., Janssens, O., Van Hoecke, S.: Eliminating noise in the matrix profile. In: Proceedings of the 8th International Conference on Pattern Recognition Applications and Methods, pp. 83–93, February 2019
5. De Paepe, D., et al.: A generalized matrix profile framework with support for contextual series analysis. Eng. Appl. Artif. Intell. (2019, accepted)
6. Silva, D.F., Batista, G.E.: Elastic time series motifs and discords. In: 17th IEEE International Conference on Machine Learning and Applications (ICMLA), pp. 237–242 (2018)
7. Gharghabi, S., Ding, Y., Yeh, C.C.M., Kamgar, K., Ulanova, L., Keogh, E.: Matrix profile VIII: domain agnostic online semantic segmentation at superhuman performance levels. In: 2017 IEEE International Conference on Data Mining (ICDM), pp. 117–126. IEEE, November 2017
8. Gharghabi, S., Imani, S., Bagnall, A., Darvishzadeh, A., Keogh, E.: MPdist: a novel time series distance measure to allow data mining in more challenging scenarios. In: IEEE International Conference on Data Mining (ICDM), pp. 965–970 (2018)
9. Imani, S., Madrid, F., Ding, W., Crouter, S., Keogh, E.: Time series snippets: a new primitive for time series data mining. In: IEEE International Conference on Big Knowledge (ICBK), pp. 382–389. IEEE (2018)
10. Keogh, E., Kasetty, S.: On the need for time series data mining benchmarks. In: Proceedings of the 8th ACM SIGKDD International Conference on Knowledge Discovery and Data Mining, p. 102 (2002)
11. Lavin, A., Ahmad, S.: Evaluating real-time anomaly detection algorithms - the numenta anomaly benchmark. In: IEEE 14th International Conference on Machine Learning and Applications (ICMLA), pp. 38–44, December 2015
12. Linardi, M., Zhu, Y., Palpanas, T., Keogh, E.: Matrix profile X. In: Proceedings of the 2018 International Conference on Management of Data (SIGMOD), pp. 1053–1066 (2018)
13. Mueen, A., Keogh, E., Zhu, Q., Cash, S., Westover, B.: Exact discovery of time series motifs. In: Proceedings of the SIAM International Conference on Data Mining (2009)
14. Mueen, A., Viswanathan, K., Gupta, C., Keogh, E.: The fastest similarity search algorithm for time series subsequences under euclidean distance (2015)

15. Papadimitriou, S., Faloutsos, C.: Streaming pattern discovery in multiple time-series. In: International Conference on Very Large Data Bases (VLDB), pp. 697–708 (2005)
16. Rafiei, D.: On similarity-based queries for time series data. In: Proceedings 15th International Conference on Data Engineering, pp. 410–417. IEEE (1999)
17. Reiss, A., Stricker, D.: Introducing a new benchmarked dataset for activity monitoring. In: International Symposium on Wearable Computers, pp. 108–109 (2012)
18. Vitter, J.S.: Random sampling with a reservoir. ACM Trans. Math. Softw. (TOMS) 11(1), 37–57 (1985)
19. Wang, X., Lin, J., Patel, N., Braun, M.: A self-learning and online algorithm for time series anomaly detection, with application in CPU manufacturing. In: Proceedings of the 25th ACM International on Conference on Information and Knowledge Management - CIKM 2016, pp. 1823–1832. ACM Press, New York (2016)
20. Yeh, C.C.M., Kavantzas, N., Keogh, E.: Meaningful multidimensional motif discovery. In: IEEE International Conference on Data Mining (ICDM), pp. 565–574. IEEE (2017)
21. Yeh, C.C.M., Kavantzas, N., Keogh, E.: Using weakly labeled time series to predict outcomes. Proc. VLDB Endow. 10(12), 1802–1812 (2017)
22. Yeh, C.C.M., Van Herle, H., Keogh, E.: The matrix profile allows visualization of salient subsequences in massive time series. In: Proceedings of the IEEE International Conference on Data Mining, ICDM, pp. 579–588 (2017)
23. Yeh, C.C.M., et al.: All pairs similarity joins for time series: a unifying view that includes motifs, discords and shapelets. In: IEEE 16th International Conference on Data Mining (ICDM), pp. 1317–1322 (2016)
24. Zhu, Y., Imamura, M., Nikovski, D., Keogh, E.: Time series chains: a new primitive for time series data mining. In: IEEE International Conference on Data Mining (ICDM), pp. 695–704 (2017)
25. Zhu, Y., Mueen, A., Keogh, E.: Admissible Time Series Motif Discovery with Missing Data (2018). https://arxiv.org/pdf/1802.05472.pdf
26. Zhu, Y., Yeh, C.C.M., Zimmerman, Z., Kamgar, K., Keogh, E.: SCRIMP++: time series motif discovery at interactive speeds. In: IEEE International Conference on Data Mining (ICDM), pp. 837–846 (2018)
27. Zhu, Y., et al.: Exploiting a novel algorithm and GPUs to break the one hundred million barrier for time series motifs and joins. In: 2016 IEEE 16th International Conference on Data Mining (ICDM), pp. 739–748 (2016)

Enforcing the General Planar Motion Model: Bundle Adjustment for Planar Scenes

Marcus Valtonen Örnhag[1]([⊠])(iD) and Mårten Wadenbäck[2](iD)

[1] Centre for Mathematical Sciences, Lund University, Lund, Sweden
marcus.valtonen_ornhag@math.lth.se
[2] Department of Mathematical Sciences,
Chalmers University of Technology and the University of Gothenburg,
Gothenburg, Sweden
marten.wadenback@chalmers.se

Abstract. In this paper we consider the case of planar motion, where a mobile platform equipped with two cameras moves freely on a planar surface. The cameras are assumed to be directed towards the floor, as well as being connected by a rigid body motion, which constrains the relative motion of the cameras and introduces new geometric constraints. In the existing literature, there are several algorithms available to obtain planar motion compatible homographies. These methods, however, do not minimise a physically meaningful quantity, which may lead to issues when tracking the mobile platform globally. As a remedy, we propose a bundle adjustment algorithm tailored for the specific problem geometry. Due to the new constrained model, general bundle adjustment frameworks, compatible with the standard six degree of freedom model, are not directly applicable, and we propose an efficient method to reduce the computational complexity, by utilising the sparse structure of the problem. We explore the impact of different polynomial solvers on synthetic data, and highlight various trade-offs between speed and accuracy. Furthermore, on real data, the proposed method shows an improvement compared to generic methods not enforcing the general planar motion model.

Keywords: Planar motion · Bundle adjustment · SLAM · Visual Odometry

1 Introduction

The prototypical problem in geometric computer vision is the so called Structure from Motion (SfM) problem [12,24]; the objective of which is to recover the scene geometry and camera poses from a collection of images of a scene. The SfM problem has, in some form or other, been studied since the very earliest days of photography, and many fundamental aspects of SfM were well understood already by the end of the 19th century [23]. Solving SfM problems of meaningful size and with actual image data, however, has been made possible only through the computerisation efforts that were commenced in the late 1970s, and which have since led to increasingly automatic methods for SfM. Modern SfM systems, e.g. *Bundler* [22] and other systems under the wider

© Springer Nature Switzerland AG 2020
M. De Marsico et al. (Eds.): ICPRAM 2019, LNCS 11996, pp. 119–135, 2020.
https://doi.org/10.1007/978-3-030-40014-9_6

BigSFM banner[1] [1,8], have managed to produce impressive city-scale reconstructions from large unordered and unlabelled sets of images.

A major paradigm in SfM, which has proven hugely successful, is Bundle Adjustment (BA) [26], which treats SfM as a large optimisation problem. With a parameterisation describing the scene geometry and the cameras, BA employs numerical optimisation techniques to find parameter values which best explain the observed images. Here, 'best' is determined by evaluating a cost function which is often—but not always— chosen as the sum of squared geometric reprojection errors. The BA formulation of the SfM problem puts it in a unified framework which still has extensive model flexibility, e.g. with regards to (a) assumptions on the camera calibration, (b) different cost functions, and (c) different parameterisations of the cameras and the scene geometry— including implicit and explicit constraints to enforce a particular motion model.

While camera based Simultaneous Localisation and Mapping (SLAM) and Visual Odometry (VO) can be thought of as special classes of SfM, the computational effort to approach SLAM and VO via BA has traditionally been inhibiting, and for this reason, BA has mostly been used in offline batch processing systems such as the BigSFM systems mentioned earlier. During the last two decades, however, SLAM and VO systems have started incorporating regular BA steps to improve the consistency of the reconstruction and the precision of the camera pose estimation. Performance improvements across the spectrum—the algorithms, their implementation, the hardware—are paving the way for application specific BA to make its entrance in the area of real-time systems.

Especially in the case of visual SLAM, there are a number of factors which can be exploited to alleviate the computational burden compared to a more generic SfM system. The images are acquired in an ordered sequence, and this can significantly speed up the search for correspondences by avoiding the expensive 'all-vs-all' matching. Additionally, a suitable motion model may often be incorporated in a SLAM system, which can be used e.g. (a) to further speed up the search for correspondences by predicting feature locations in subsequent images [5,6], (b) to facilitate faster and more accurate local motion estimation via nonholonomic constraints [20,21,39] or other constraints which reduce the set of parameters [29,33], or (c) to enforce globally a planar motion assumption on the camera motion [10,18,32].

In this paper, we present a BA approach to visual SLAM for the case of a stereo rig, where the cameras do not necessarily have an overlapping field of view, and where each of the two cameras move in parallel to a common ground plane. The present paper is an extension of the system described earlier in [30], to which a more extensive experimental evaluation has been added. In particular, we have investigated how initialisation using planar motion compatible homographies based on minimal [33] or non-minimal [29] polynomial solvers affect the final reconstruction.

2   Related Work

Planar Motion is a frequently occurring constrained camera motion, which arises naturally when cameras are attached to a ground vehicle operating on a planar ground

---

[1] http://www.cs.cornell.edu/projects/bigsfm/.

surface. As mentioned in the introduction, deliberately enforcing planar motion can help to improve the quality of the reconstruction.

An early SfM approach to plane constrained visual navigation was proposed by Wiles and Brady [34,35]. They suggested a hierarchical framework of camera parameterisations, and explored in detail the remaining structural ambiguity for each of these. The lasting contribution of this work lies chiefly in its classification and description of the different modes of motion. The least ambiguous level in the case of planar motion—which they called α-structure—contains only an arbitrary global scaling ambiguity and an arbitrary planar Euclidean transformation parallel to the ground plane, and is precisely the level aimed at in the present paper.

If the optical axis of the camera is either orthogonal or parallel to the ground plane, the parameterisation can be much simplified compared to the general case described by Wiles and Brady. This situation can of course also be achieved if the camera tilt is known with sufficient precision to allow a transformation to, e.g., an overhead view. An approach for this case by Ortín and Montiel parameterises the essential matrix explicitly in the motion parameters, and then estimates the parameters using either a linear three-point method or a non-linear two-point method [18]. Scaramuzza used essentially the same parameterisation of the essential matrix, but combined it with an additional nonholonomic constraint based on the assumption that the local motion is a circular motion [20,21]. Because of this additional constraint, the local motion can be computed from only one point correspondence, and this allows for an exceptionally efficient outlier removal scheme based on histogram voting.

Since the essential matrix is a homogeneous entity, it does not capture the length of the translation, and the maintaining of a consistent global scale then requires some additional information. One possibility for this, explored by Chen and Liu, is to add a second camera [4]. This allows the length of the local translation to be computed in terms of the distance between the two cameras, and since this remains constant, it provides a way to prevent scale drift.

If the camera is oriented such that it views a reasonable part of the ground plane, an alternative to using the essential matrix is to instead use homographies for the local motion estimation. This has the advantage that the length of the translation between frames can be expressed in terms of the height above the ground plane, which thus defines the global scale. The homography based approach by Liang and Pears is based on an eigendecomposition of the homography matrix, and it is shown that the rotation about the vertical axis can be determined from the eigenvalues, regardless of the camera tilt [14]. Hajjdiab and Laganière parameterised the homography matrix under the assumption of only one tilt angle, and then transformed the images into a synthetic overhead view to compute the residual rigid body motion in the plane [10].

A more recent homography based method by Wadenbäck and Heyden, which also exploits a decoupling of the camera tilt and the camera motion, uses an alternating iterative estimation scheme to compute the two tilt angles and the three motion parameters [31,32]. Zienkiewicz and Davison solved the same 5-DoF problem through a joint non-linear optimisation over all five parameters to achieve a dense matching of successive views, with the implementation running on a GPU to reach very high frame rates [39].

Valtonen Örnhag and Heyden extended the general 5-DoF situation to handle a binocular setup, where the two cameras are connected by a fixed (but unknown) rigid body motion in 3D, and where the fields of view do not necessarily overlap [27,28].

Bundle Adjustment is used to optimise a set of structure and motion parameters, and is typically performed over several camera views. Triggs et al. give an excellent overview [26]. Since the number of parameters optimised over is in most cases very large, naïve implementations will not work, and care must be taken to exploit the problem structure (e.g. the sparsity pattern of the Jacobian).

Generic software packages for bundle adjustment, which use sparsity of the Jacobian matrix together with Schur complementation to speed up the computations, include *SBA* (Sparse Bundle Adjustment) by Lourakis and Argyros, *sSBA* (Sparse Sparse Bundle Adjustment) by Konolige, and *SSBA* (Simple Sparse Bundle Adjustment) by Zach [13,16,37].

Additional performance gains may sometimes be obtained through parallelisation. GPU accelerated BA systems using parallelised versions of the Levenberg–Marquardt algorithm [11] and the conjugate gradients method [36] have been presented e.g. by Hänsch et al. and by Wu et al.. More recently, distributed approaches by e.g. Eriksson et al. and by Zhang et al. have employed splitting methods to make very large SfM problems tractable [7,38].

The present paper extends the sparse bundle adjustment system for the binocular planar motion case by Valtonen Örnhag and Wadenbäck. The aim of our approach is to exploit the particular structure in the Jacobian which arises due to the planar motion assumption for the two cameras. We demonstrate how this particular situation can be attacked via the use of nested Schur complementations when solving the normal equations. In comparison to the earlier paper [30], we have significantly extended the experimental evaluation of the system. Additionally, we have investigated the effect of enforcing the planar motion assumption earlier on a local level, by using homographies estimated such that they are compatible with this assumption [29,33].

3   Theory

3.1   Problem Geometry

The geometrical situation we consider in this paper is that of two cameras which have been rigidly mounted onto a mobile platform. Due to this setup, which is illustrated in Fig. 1, the cameras are connected by a rigid body motion which remains constant over time but which is initially not known. Each camera is assumed to be mounted in such a way that it can view a portion of the ground plane, but it is *not* a requirement that the cameras have any portion of their fields of view in common. The world coordinate system is chosen such that the ground plane is positioned at $z = 0$, whereas the cameras move in the planes $z = a$ and $z = b$, respectively. We may also, without loss of generality, assume that the centre of rotation of the mobile platform coincides with the centre of the first camera.

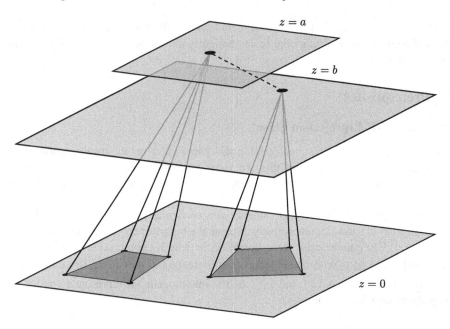

Fig. 1. Illustration of the problem geometry considered in this paper. Two cameras are assumed to be rigidly mounted on a mobile platform, and may be positioned at different heights above, the ground floor, hence move in the planes $z = a$ and $z = b$. Due to the rigidity assumption, the relative orientation between them are constant, and so is the overhead tilt. Figure reproduced from [30].

3.2 Camera Parameterisation

We shall adopt the camera parameterisation for internally calibrated monocular planar motion that was introduced in [31]. With this parameterisation, the camera matrix associated with the image taken at position j will be

$$P^{(j)} = R_{\psi\theta} R_{\varphi}^{(j)} [I \mid -t^{(j)}], \tag{1}$$

where $R_{\psi\theta}$ is a rotation θ about the y-axis followed by a rotation of ψ about the x-axis. The motion of the mobile platform contains for each frame a rotation $\varphi^{(j)}$ about the z-axis, encoded as $R_{\varphi}^{(j)}$, and a vector $t^{(j)}$ for the translational part. The second camera, which is related to the first camera through a constant rigid body motion, uses the parameterisation

$$P'^{(j)} = R_{\psi'\theta'} R_{\eta} T_{\tau}(b) R_{\varphi}^{(j)} [I \mid -t^{(j)}], \tag{2}$$

introduced in [27]. Here, ψ' and θ' are the tilt angles (defined in the same way as for the first camera), τ is the relative translation between the camera centres and η is the constant rotation about the z-axis relative to the first camera. We do not assume any prior knowledge of these constant parameters. Define the translation matrix $T_{\tau}(b)$ as

$T_\tau(b) = I - \tau n^\mathsf{T}/b$, where $\tau = (\tau_x, \tau_y, 0)^\mathsf{T}$, n is a floor normal and b is the height above the ground floor. The global scale ambiguity allows us to set $a = 1$ without any loss of generality.

4 Prerequisites

4.1 Geometric Reprojection Error

The particular BA problem considered in this paper concerns the minimisation of the *geometric reprojection error* in the two views over the entire motion sequence. In order to write down this cost function explicitly we need to introduce some additional notation.

For this purpose, let the two cameras at a particular position j be given by the expressions in (1) and (2), respectively. We use the homogeneous representation $X_i = (X_i, Y_i, 0, 1)^\mathsf{T}$ to parameterise the estimate of the i:th 3D point, corresponding to the measured image point with inhomogeneous representations $x_i^{(j)}$ in the first camera and $x_i'^{(j)}$ in the second. Let $\hat{\bar{x}}_i^{(j)}$ and $\hat{\bar{x}}_i'^{(j)}$ be the inhomogeneous representations for the projections into the two views, i.e.

$$\begin{bmatrix} \hat{\bar{x}}_i^{(j)} \\ 1 \end{bmatrix} \sim P^{(j)} X_i \quad \text{and} \quad \begin{bmatrix} \hat{\bar{x}}_i'^{(j)} \\ 1 \end{bmatrix} \sim P'^{(j)} X_i. \tag{3}$$

Given N stereo camera locations and M scene points, the geometric reprojection error that we seek to minimise can now be written concisely as

$$E(\beta) = \sum_{i=1}^{N} \sum_{j=1}^{M} \|r_{ij}\|_2^2 + \|r_{ij}'\|_2^2, \tag{4}$$

where β is the parameter vector consisting of the camera parameters and the scene point parameters, and where r_{ij} and r_{ij}' are the residuals

$$r_{ij} = x_i^{(j)} - \hat{\bar{x}}_i^{(j)} \quad \text{and} \quad r_{ij}' = x_i'^{(j)} - \hat{\bar{x}}_i'^{(j)}. \tag{5}$$

4.2 The Levenberg–Marquardt Algorithm

We will in this approach use the Levenberg–Marquardt algorithm (LM) when minimising (4). There are of course other alternatives to the LM algorithm, e.g. the dog-leg solver [15] and preconditioned CG [3]; however, LM is one of the most commonly used algorithms for BA, and is used in major modern systems such as SBA [16] and sSBA [13]. Note that these systems do not account for the particular problem geometry that we consider in this paper, which forces some extrinsic parameters to be shared among all camera matrices.

We will not go into details of the LM algorithm here—please refer to more extensive treatments in e.g. [26] and [16] for a more complete discussion—but for future reference we simply recall that it works by iteratively solving the augmented normal equations

$$(J^\mathsf{T} J + \mu I)\, \delta = J^\mathsf{T} \varepsilon \tag{6}$$

until some convergence criteria have been met. Here J is the Jacobian associated with the cost function (4), ε is the residual vector, and $\mu \geq 0$ is the iteratively adjusted *damping parameter* of the LM algorithm.

4.3   Obtaining an Initial Solution for the Camera Parameters

Homographies can be estimated in a number of different ways; however, the classical approach is to compute point correspondences from matching robust feature points in subsequent images. Popular feature extraction algorithms include SIFT [17] and SURF [2], but many more are available and implemented in various computer vision software. When the putative point correspondences have been matched a popular choice is to use RANSAC (or similar frameworks) to robustly estimate a homography. Such an approach is suitable in order to discard mismatched feature points. A well-known method is the Direct Linear Transform (DLT); however, it requires four point correspondences, and does not generate a homography compatible with the general planar motion model. A good rule of thumb is to use a minimal amount of point correspondences, since the probability of finding a set of points containing only inliers decreases with each additional point that is used. However, as e.g. Pham et al. point out, for very severely noisy data it may in some cases still be preferable to use a non-minimal set [19].

In [33] a minimal solver compatible with the general planar motion model was studied. It was shown that a homography compatible with the general planar motion model must fulfil 11 quartic constraints, and that, a minimal solver only requires 2.5 point correspondences. In a recent paper, a variety of different non-minimal polynomial solvers are considered, partly because of execution time, but also because of sensitivity to noise [29]. These non-minimal solvers enforce a subset of the necessary and sufficient conditions for compatibility with the general planar motion model, thus enforcing a weaker form of it. By accurately making a trade-off between fitting the model constraints (i.e. using more model constraints) and tuning to data (i.e. using more point correspondences), one can increase the performance for noisy data. It is important to note that the assumption of constant tilt parameters cannot be enforced by only considering a single homography, and, therefore, pre-optimisation in an early step of the complete SfM pipeline is not guaranteed to yield better performance.

Once the homographies are obtained, one may enforce the constant tilt constraint by employing the method proposed by Wadenbäck and Heyden [32], to obtain a good initial solution for the monocular case. The method starts by computing the overhead tilt $R_{\psi\theta}$ from an arbitrary number of homographies, followed by estimating the translation and orientation about the floor normal.

The method by Valtonen Örnhag and Heyden [27] extended the method to include the stereo case, and starts off by treating the two stereo trajectories individually, and estimates the tilt parameters by employing the monocular method described in the previous paragraph. Once the monocular parameters are known for the individual tracks, the relative pose can be extracted by minimising an algebraic error in the relative translation between the cameras, followed by estimating the relative orientation about the floor normal.

4.4 Obtaining an Initial Solution for the Scene Points

Linear triangulation of scene points does not guarantee that all points lie in a plane, and the resulting initial solution would not be compatible with the general planar motion model. In order to obtain a physically meaningful solution we make use of the fact that there is a homography relating the measured points and the ground plane positioned at $z = 0$.

Given a camera P, an image point x and the corresponding scene point $X \sim (X, Y, 0, 1)^\mathsf{T}$, they are related by $x \sim PX = H\tilde{X}$, where H is the sought homography. By denoting the i:th column of P by P_i, it may be expressed as $H = [P_1\ P_2\ P_4]$, where $\tilde{X} \sim (X, Y, 1)^\mathsf{T}$ contains the unknown scene point coordinates. It follows that the corresponding scene point can be extracted from $\tilde{X} \sim H^{-1}x$.

In the presence of noise, using more than one camera results in different scene points, which all will be projected onto the plane $z = 0$. In order to triangulate the points we compute the centre mass; such an approach is computationally inexpensive, however, it is not robust to outliers, which have to be excluded in order to get a reliable result.

5 Planar Motion Bundle Adjustment

5.1 Block Structure of the Jacobian

Denote the unknown and constant parameters for the first camera path by $\gamma = (\psi, \theta)$ and the second camera path by $\gamma' = (\psi', \theta', \tau_x, \tau_y, b, \eta)$. Furthermore, let the nonconstant parameters for position j be denoted by $\xi_j = (\varphi^{(j)}, t_x^{(j)}, t_y^{(j)})$. Given N stereo camera positions and M scene points, the following, highly structured Jacobian J, is obtained

$$
J = \begin{bmatrix}
\Gamma_{11} & A_{11} & & & & B_{11} & & \\
\vdots & & \ddots & & & & \vdots & \\
\Gamma_{1N} & & & A_{1N} & & B_{1N} & \\
\vdots & \vdots & \vdots & \vdots & & & & \ddots \\
\Gamma_{M1} & A_{M1} & & & & & B_{M1} \\
\vdots & & \ddots & & & & & \vdots \\
\Gamma_{MN} & & & A_{MN} & & & & B_{MN} \\
\Gamma'_{11} & A'_{11} & & & & B'_{11} & & \\
\vdots & & \ddots & & & & \vdots & \\
\Gamma'_{1N} & & & A'_{1N} & & B'_{1N} & \\
\vdots & \vdots & \vdots & \vdots & & & & \ddots \\
\Gamma'_{M1} & A'_{M1} & & & & & B'_{M1} \\
\vdots & & \ddots & & & & & \vdots \\
\Gamma'_{MN} & & & A'_{MN} & & & & B'_{MN}
\end{bmatrix}, \tag{7}
$$

where we use the following notation for the derivative blocks

$$
\begin{aligned}
A_{ij} &= \frac{\partial r_{ij}}{\partial \xi_j}, & B_{ij} &= \frac{\partial r_{ij}}{\partial \tilde{X}_i}, & \Gamma_{ij} &= \frac{\partial r_{ij}}{\partial \gamma}, \\
A'_{ij} &= \frac{\partial r'_{ij}}{\partial \xi_j}, & B'_{ij} &= \frac{\partial r'_{ij}}{\partial \tilde{X}_i}, & \Gamma'_{ij} &= \frac{\partial r'_{ij}}{\partial \gamma'},
\end{aligned} \tag{8}
$$

where $\tilde{X}_i = (X_i, Y_i)$ are the unknown scene coordinates. This can be written in a more compact manner as

$$J = \begin{bmatrix} \Gamma & 0 & A & B \\ 0 & \Gamma' & A' & B' \end{bmatrix}.$$
(9)

5.2 Utilising the Sparse Structure

In SfM, the number of scene points is often significantly larger than the number of cameras, which makes Schur complementation tractable, and can significantly decrease the execution time. Standard Schur complementation is, however, not directly applicable due to the constant parameters giving rise to the blocks Γ and Γ'. We will, however, show in this section, that it is indeed possible to use *nested Schur complements*, i.e. to recursively apply Schur complements to different parts, and that, in fact, several of the intermediate computations can be stored, thus drastically decreasing the computational time. First, note that the approximate Hessian $J^\mathsf{T}J$, in compact form, can be written

$$J^\mathsf{T}J = \begin{bmatrix} C & E \\ E^\mathsf{T} & D \end{bmatrix}.$$
(10)

Here the contribution from the constant parameters are stored in C, the contribution from the nonconstant parameters and the scene points are stored in D, and the mixed contributions are stored in E. Furthermore, the matrix D can be written as

$$D = \begin{bmatrix} U & W \\ W^\mathsf{T} & V \end{bmatrix},$$
(11)

with block diagonal matrices $U = \operatorname{diag}(U_1, \ldots, U_N)$ and $V = \operatorname{diag}(V_1, \ldots, V_M)$, where

$$\begin{aligned}
U_j &= \sum_{i=1}^{M} A_{ij}^\mathsf{T} A_{ij} + A_{ij}'^\mathsf{T} A_{ij}', \\
V_i &= \sum_{j=1}^{N} B_{ij}^\mathsf{T} B_{ij} + B_{ij}'^\mathsf{T} B_{ij}', \\
W_{ij} &= A_{ij}^\mathsf{T} B_{ij} + A_{ij}'^\mathsf{T} B_{ij}'.
\end{aligned}$$
(12)

First, note that the system $(D + \mu I)\delta = \varepsilon$, where D is defined as in (11), is not affected by the constant parameters. Such a system reduces to that of the unconstrained case, which can be solved using standard SfM frameworks, such as SBA, or other packages utilising Schur complementation.

We will now show how to efficiently treat the decomposition of (10) as nested Schur complements, by reducing the problem to a series of subproblems of the form used in SBA and other computer vision software packages. In order to do so, consider the augmented normal equations (6) in block form

$$\begin{bmatrix} C^* & E \\ E^\mathsf{T} & D^* \end{bmatrix} \begin{bmatrix} \delta_c \\ \delta_d \end{bmatrix} = \begin{bmatrix} \varepsilon_c \\ \varepsilon_d \end{bmatrix},$$
(13)

where $C^* = C + \mu I$ and $D^* = D + \mu I$ denote the augmented matrices, with the added contribution from the damping factor μ, as in (6). Now, utilising Schur complementation yields

$$\begin{bmatrix} C^* - ED^{*-1}E^\mathsf{T} & 0 \\ E^\mathsf{T} & D^* \end{bmatrix} \begin{bmatrix} \delta_c \\ \delta_d \end{bmatrix} = \begin{bmatrix} \varepsilon_c - ED^{*-1}\varepsilon_d \\ \varepsilon_d \end{bmatrix}. \tag{14}$$

Let us take a step back and reflect over the consequences of the above equation. First, note that D^{*-1} is present in (14) twice, and is infeasible to compute explicitly. This can be avoided by introducing the auxiliary variable δ_{aux}, defined as

$$D^* \delta_{\text{aux}} = \varepsilon_d. \tag{15}$$

Again, such as system is not affected by the constraints of the constant parameters, and can be solved with standard computer vision software. Furthermore, we may introduce Δ_{aux} and solve the system $D^* \Delta_{\text{aux}} = E^\mathsf{T}$ in a similar manner by iterating over the columns of E^T. Since the number of constant parameters are low, such an approach is highly feasible, but the performance can be further boosted by storing the Schur complement and the intermediate matrices not depending on the right-hand side, from the previous computations of obtaining δ_{aux} from (15).

When the auxiliary variables have been obtained, we proceed to compute δ_c from

$$(C^* - E\Delta_{\text{aux}}) \delta_c = \varepsilon_c - E\delta_{\text{aux}}, \tag{16}$$

and, lastly, δ_d by back-substitution

$$D^* \delta_d = \varepsilon_d - E^\mathsf{T}\delta_c. \tag{17}$$

Again, by storing the computation of the Schur complement and intermediate matrices, these can be reused to solve (17) efficiently.

6   Experiments

6.1   Initial Solution

The inter-image homographies were estimated using the MSAC algorithm [25] from point correspondences by extracting SURF keypoints and applying a KNN algorithm to establish the matches. In the first experiment, we use the standard DLT solver, the minimal 2.5 pt solver [33] and the four different polynomial solvers studied in [29].

In all experiments we use all available homographies, and extract the monocular parameters using the method proposed in [32]. Similarly, the binocular parameters were extracted using [27]. When all motion parameter have been estimated the camera path is reconstructed by aligning the first camera position to the origin, and use the estimated camera poses to triangulate the scene points as in Sect. 4.4.

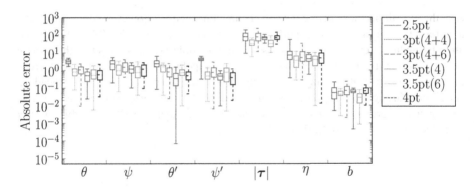

Fig. 2. Errors before applying BA. The angles are measured in degrees, and the translation in pixels.

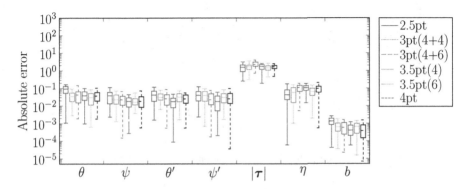

Fig. 3. Errors after applying BA. The angles are measured in degrees, and the translation in pixels.

6.2 Impact of Pre-processing Steps

In this section we work with synthetic data in order to have access to accurate ground truth data. We generate an image sequence from a high-resolution image, depicting a floor, which is the typical use case for the algorithm. This is done by constructing a path compatible with the general planar motion model, and project that part of the floor through the camera and extract the corresponding image. The resulting image is 400×400 pixels, and all cameras are set to a field of view of $90°$, with parameters $\psi = -2°$, $\theta = -4°$, $\psi' = 6°$, $\theta' = 4°$, $\tau = (0\ 400)$, $\eta = 20°$ and $b = 1$. In total, the image sequence consists of 20 images. Lastly, to simulate image noise, we add Gaussian noise with a standard deviation of five pixels, where the pixel depth allows 256 different intensities per channel.

In order to study the difference in accuracy for the constant parameters, we proceed by obtaining homographies as described in Sect. 6.1, using the minimal 2.5 point solver [33], four non-minimal solvers [29] and the DLT equations (4 point). The accuracy, over 50 iterations, is reported before BA, in Fig. 2, and after BA, in Fig. 3. In general, the overall performance of the solvers are almost equal; however, some tendencies are present. The minimal solver performs worse than the other before BA, but this deviation is smaller after BA, although present. One possible explanation is that the general planar motion model is enforced too early in the pipeline—in fact, since it is enforced between two consecutive image pairs only, it does not guarantee that the overhead tilt is constant throughout the entire sequence, and thus, in the presence of noise, the error propagates differently, compared to the other methods that partially (non-minimal) or completely (DLT) tune to the data.

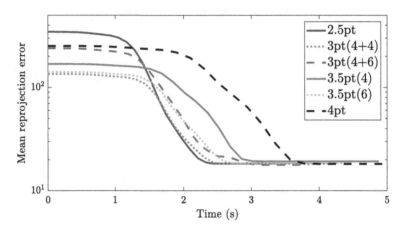

Fig. 4. Mean reprojection error vs execution time (s) over 50 iterations.

Overall, the performance is acceptable after BA, regardless of how the homographies are obtained. Hence, the differentiating factors come down to convergence rates. For the same problem instances as in the previous section we also save the convergence history in terms of the mean reprojection error and the execution time in seconds. The results are shown in Fig. 4. It is clear that the execution time for reaching convergence increase with the number of point correspondences required by the polynomial solvers. This suggests that one can make a trade-off between speed and accuracy when designing a planar motion compatible BA framework by choosing different solvers, in order to suit ones specific needs. Note, however, that the implementation used in this paper is a native Matlab implementation, and that the absolute timings can be greatly improved by careful implementation; however, the relative execution time between the solvers will be similar.

6.3   Bundle Adjustment Comparison

In this section we compare the qualitative difference between enforcing the general planar motion model versus the general unconstrained six degree of freedom model on a real dataset. Currently, there is not a good or well-established dataset compatible with the general planar motion model, and as a substitute, we use the KITTI Visual Odometry/SLAM benchmark [9]. Since many sequences or subsequences depict urban environments with paved roads, the general planar motion model can roughly be applied. In case of clear violation of the general planar motion model, we proceed to use only subsequences where the model is applicable. As we are only interested by the road in front of the vehicle, and not the sky and other objects by the roadside, we proceed to crop a part of the image prior to estimating the homography. An example of this is shown in Fig. 5.

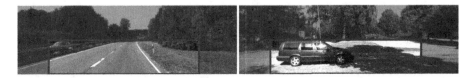

Fig. 5. Images from the KITTI Visual Odometry/SLAM benchmark, Sequence 01 (left) and 03 (right). Since the algorithm is homography-based the images are cropped *a priori* in order to contain a significant portion of planar or near planar surface. Such an assumption is not valid on all sequences of the dataset, however, certain cases, such as the highway of Sequence 01 (left) is a good candidate. There are several examples where occlusions occur, such as the car in Sequence 03 (right). These situations typically occur at crossroads and turns. Image credit: KITTI dataset [9].

We use SBA [16] to enforce the general 6-DoF model from the initial trajectory obtained using the traditional 4-point DLT solver, and from the same trajectory our proposed BA algorithm is used. The same thresholds for absolute and relative errors, termination control and damping factors are used for both methods. Furthermore, we do not match features between the stereo views, in order to demonstrate that enforcing the model is enough to increase the overall performance. The results are shown in Fig. 6.

In most cases it is favourable to impose the proposed method compared to the general 6-DoF method, using SBA. Furthermore, note that irregularities that are present in the initial trajectory is often transferred to the solutions obtained by SBA, thus producing physically improbable solutions. These irregularities are rarely seen using the proposed method, which results in smooth realistic trajectories under general conditions, regardless of whether the initial solution contains irregularities or not.

In fact, it is interesting to see what happens in cases where the general planar motion model is violated. Such an instance occurs in Fig. 6(b) depicting Sequence 03, and is due to the car approaching a crossroads, where a passing vehicle enters the field of view. The observed car, and the surroundings, are highly non-planar; one would, perhaps, expect such a clear violation to result in completely unreliable output, however, the only inconsistency in comparison to the ground truth, is that the resulting turn is too sharp, and the remaining path is consistent with the ground truth. This is not true for the general 6-DoF model, where several obvious inconsistencies are present.

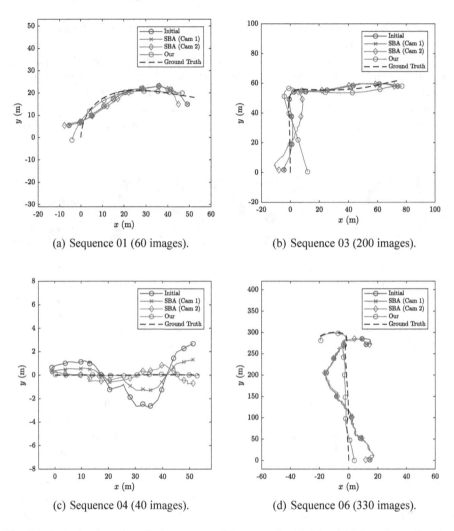

(a) Sequence 01 (60 images).

(b) Sequence 03 (200 images).

(c) Sequence 04 (40 images).

(d) Sequence 06 (330 images).

Fig. 6. Estimated trajectories of subsequences of Sequence 01, 03, 04 and 06. In order to align the estimated paths with the ground truth, Procrustes analysis has been carried out. N.B. the different aspect ratio in (c), which is intentionally added in order to clearly visualise the difference. Figure reproduced from [30].

7   Conclusion

In this paper a novel bundle adjustment method has been devised, which enforces the general planar motion model. We provide an efficient implementation scheme that exploits the sparse structure of the Jacobian, and, additionally, avoids recomputing unnecessary quantities, making it highly attractive for real-time computations.

The performance of different polynomial solvers are studied, in terms of both accuracy and speed, taking the entire bundle adjustment framework into account. We discuss how enforcing different polynomial constraints, through planar motion compatible homography solvers, in an early part of the bundle adjustment framework affect the end results. Furthermore, we discuss which trade-offs between speed and accuracy that can be made to suit ones specific priorities.

The proposed method has been tested on real data and was compared to state-of-the-art methods for sparse bundle adjustment, for which it performs well, and gives physically accurate solutions, despite some model assumptions not being fulfilled.

Acknowledgements. This work has been funded by the Swedish Research Council through grant no. 2015-05639 'Visual SLAM based on Planar Homographies'.

References

1. Agarwal, S., Furukawa, Y., Snavely, N., Simon, I., Curless, B., Seitz, S.M., Szeliski, R.: Building Rome in a day. Commun. ACM **54**(10), 105–112 (2011)
2. Bay, H., Tuytelaars, T., Van Gool, L.: SURF: speeded up robust features. In: Leonardis, A., Bischof, H., Pinz, A. (eds.) ECCV 2006. LNCS, vol. 3951, pp. 404–417. Springer, Heidelberg (2006). https://doi.org/10.1007/11744023_32
3. Byröd, M., Åström, K.: Conjugate gradient bundle adjustment. In: Daniilidis, K., Maragos, P., Paragios, N. (eds.) ECCV 2010. LNCS, vol. 6312, pp. 114–127. Springer, Heidelberg (2010). https://doi.org/10.1007/978-3-642-15552-9_9
4. Chen, T., Liu, Y.H.: A robust approach for structure from planar motion by stereo image sequences. Mach. Vis. Appl. (MVA) **17**(3), 197–209 (2006)
5. Davison, A.J.: Real-time simultaneous localisation and mapping with a single camera. In: International Conference on Computer Vision (ICCV), Nice, France, pp. 1403–1410, October 2003
6. Davison, A.J., Reid, I.D., Molton, N.D., Stasse, O.: MonoSLAM: real-time single camera SLAM. IEEE Trans. Pattern Anal. Mach. Intell. (PAMI) **29**(6), 1052–1067 (2007)
7. Eriksson, A., Bastian, J., Chin, T., Isaksson, M.: A consensus-based framework for distributed bundle adjustment. In: Conference on Computer Vision and Pattern Recognition (CVPR), Las Vegas, NV, USA, pp. 1754–1762, June 2016
8. Frahm, J.-M., et al.: Building Rome on a cloudless day. In: Daniilidis, K., Maragos, P., Paragios, N. (eds.) ECCV 2010. LNCS, vol. 6314, pp. 368–381. Springer, Heidelberg (2010). https://doi.org/10.1007/978-3-642-15561-1_27
9. Geiger, A., Lenz, P., Urtasun, R.: Are we ready for autonomous driving? The KITTI vision benchmark suite. In: Conference on Computer Vision and Pattern Recognition (CVPR), Providence, RI, USA, June 2012
10. Hajjdiab, H., Laganière, R.: Vision-based multi-robot simultaneous localization and mapping. In: Canadian Conference on Computer and Robot Vision (CRV), London, ON, Canada, pp. 155–162, May 2004
11. Hänsch, R., Drude, I., Hellwich, O.: Modern methods of bundle adjustment on the GPU. In: ISPRS Annals of Photogrammetry, Remote Sensing and Spatial Information Sciences (ISPRS Congress), Prague, Czech Republic, pp. 43–50, July 2016
12. Hartley, R.I., Zisserman, A.: Multiple View Geometry in Computer Vision, 2nd edn. Cambridge University Press, Cambridge (2004)
13. Konolige, K.: Sparse sparse bundle adjustment. In: British Machine Vision Conference (BMVC), Aberystwyth, Wales, pp. 102.1–11, August 2010

14. Liang, B., Pears, N.: Visual navigation using planar homographies. In: International Conference on Robotics and Automation (ICRA), Washington, DC, USA, pp. 205–210, May 2002

15. Lourakis, M.I.A., Argyros, A.A.: Is Levenberg-Marquardt the most efficient optimization algorithm for implementing bundle adjustment? In: International Conference on Computer Vision (ICCV), Beijing, China, PRC, pp. 1526–1531, October 2005

16. Lourakis, M.I.A., Argyros, A.A.: SBA: a software package for generic sparse bundle adjustment. ACM Trans. Math. Softw. (TOMS) 36(1), 2:1–2:30 (2009)

17. Lowe, D.G.: Distinctive image features from scale-invariant keypoints. Int. J. Comput. Vis. (IJCV) 60(2), 91–110 (2004)

18. Ortín, D., Montiel, J.M.M.: Indoor robot motion based on monocular images. Robotica 19(3), 331–342 (2001)

19. Pham, T.T., Chin, T.J., Yu, J., Suter, D.: The random cluster model for robust geometric fitting. IEEE Trans. Pattern Anal. Mach. Intell. (PAMI) 36(8), 1658–1671 (2014)

20. Scaramuzza, D.: 1-point-RANSAC structure from motion for vehicle-mounted cameras by exploiting non-holonomic constraints. Int. J. Comput. Vis. (IJCV) 95(1), 74–85 (2011)

21. Scaramuzza, D.: Performance evaluation of 1-point-RANSAC visual odometry. J. Field Robot. (JFR) 28(5), 792–811 (2011)

22. Snavely, N., Seitz, S.M., Szeliski, R.: Modeling the world from internet photo collections. Int. J. Comput. Vis. (IJCV) 80(2), 189–210 (2008)

23. Sturm, P.: A historical survey of geometric computer vision. In: Real, P., Diaz-Pernil, D., Molina-Abril, H., Berciano, A., Kropatsch, W. (eds.) CAIP 2011. LNCS, vol. 6854, pp. 1–8. Springer, Heidelberg (2011). https://doi.org/10.1007/978-3-642-23672-3_1

24. Szeliski, R.: Computer Vision: Applications and Algorithms. Springer, London (2011). https://doi.org/10.1007/978-1-84882-935-0

25. Torr, P.H.S., Zisserman, A.: MLESAC: a new robust estimator with application to estimating image geometry. Comput. Vis. Image Underst. (CVIU) 78(1), 138–156 (2000)

26. Triggs, B., McLauchlan, P.F., Hartley, R.I., Fitzgibbon, A.W.: Bundle adjustment — a modern synthesis. In: Triggs, B., Zisserman, A., Szeliski, R. (eds.) IWVA 1999. LNCS, vol. 1883, pp. 298–372. Springer, Heidelberg (2000). https://doi.org/10.1007/3-540-44480-7_21

27. Valtonen Örnhag, M., Heyden, A.: Generalization of parameter recovery in binocular vision for a planar scene. In: International Conference on Pattern Recognition and Artificial Intelligence, Montréal, Canada, pp. 37–42, May 2018

28. Valtonen Örnhag, M., Heyden, A.: Relative pose estimation in binocular vision for a planar scene using inter-image homographies. In: International Conference on Pattern Recognition Applications and Methods (ICPRAM), Funchal, Madeira, Portugal, pp. 568–575, January 2018

29. Valtonen Örnhag, M.: Fast non-minimal solvers for planar motion compatible homographies. In: Proceedings of the 8th International Conference on Pattern Recognition Applications and Methods (ICPRAM), pp. 40–51. SCITEPRESS, Prague, February 2019

30. Valtonen Örnhag, M., Wadenbäck, M.: Planar motion bundle adjustment. In: Proceedings of the 8th International Conference on Pattern Recognition Applications and Methods (ICPRAM), pp. 24–31. SCITEPRESS, Prague, February 2019

31. Wadenbäck, M., Heyden, A.: Planar motion and hand-eye calibration using inter-image homographies from a planar scene. In: International Conference on Computer Vision Theory and Applications (VISAPP), Barcelona, Spain, pp. 164–168, February 2013

32. Wadenbäck, M., Heyden, A.: Ego-motion recovery and robust tilt estimation for planar motion using several homographies. In: International Conference on Computer Vision Theory and Applications (VISAPP), Lisbon, Portugal, pp. 635–639, January 2014

33. Wadenbäck, M., Åström, K., Heyden, A.: Recovering planar motion from homographies obtained using a 2.5-point solver for a polynomial system. In: 2016 IEEE International Conference on Image Processing (ICIP), pp. 2966–2970. IEEE-Institute of Electrical and Electronics Engineers Inc., September 2016. https://doi.org/10.1109/ICIP.2016.7532903

34. Wiles, C., Brady, M.: Closing the loop on multiple motions. In: Proceedings of the Fifth IEEE International Conference on Computer Vision (ICCV), pp. 308–313. IEEE Computer Society, Cambridge, June 1995

35. Wiles, C., Brady, M.: Ground plane motion camera models. In: Buxton, B., Cipolla, R. (eds.) ECCV 1996. LNCS, vol. 1065, pp. 238–247. Springer, Heidelberg (1996). https://doi.org/10.1007/3-540-61123-1_143

36. Wu, C., Agarwal, S., Curless, B., Seitz, S.M.: Multicore bundle adjustment. In: Computer Vision and Pattern Recognition (CVPR), Providence, RI, USA, pp. 3057–3064, June 2011

37. Zach, C.: Robust bundle adjustment revisited. In: Fleet, D., Pajdla, T., Schiele, B., Tuytelaars, T. (eds.) ECCV 2014. LNCS, vol. 8693, pp. 772–787. Springer, Cham (2014). https://doi.org/10.1007/978-3-319-10602-1_50

38. Zhang, R., Zhu, S., Fang, T., Quan, L.: Distributed very large scale bundle adjustment by global camera consensus. In: International Conference on Computer Vision (ICCV), Venice, Italy, pp. 29–38, October 2017

39. Zienkiewicz, J., Davison, A.J.: Extrinsics autocalibration for dense planar visual odometry. J. Field Robot. (JFR) 32(5), 803–825 (2015)

Deep Multi-biometric Fusion for Audio-Visual User Re-Identification and Verification

Mirko Marras[1], Pedro A. Marín-Reyes[2(✉)], Javier Lorenzo-Navarro[2],
Modesto Castrillón-Santana[2], and Gianni Fenu[1]

[1] Department of Mathematics and Computer Science, University of Cagliari,
V. Ospedale 72, 09124 Cagliari, Italy
{mirko.marras,fenu}@unica.it
[2] Instituto Universitario Sistemas Inteligentes y Aplicaciones Numericas
en Ingenieria (SIANI), Universidad de Las Palmas de Gran Canaria,
Campus Universitario de Tafira, 35017 Las Palmas de Gran Canaria, Spain
pedro.marin102@alu.ulpgc.es, {javier.lorenzo,modesto.castrillon}@ulpgc.es

Abstract. From border controls to personal devices, from online exam proctoring to human-robot interaction, biometric technologies are empowering individuals and organizations with convenient and secure authentication and identification services. However, most biometric systems leverage only a single modality, and may face challenges related to acquisition distance, environmental conditions, data quality, and computational resources. Combining evidence from multiple sources at a certain level (e.g., sensor, feature, score, or decision) of the recognition pipeline may mitigate some limitations of the common uni-biometric systems. Such a fusion has been rarely investigated at *intermediate level*, i.e., when uni-biometric model parameters are jointly optimized during training. In this chapter, we propose a multi-biometric model training strategy that digests face and voice traits in parallel, and we explore how it helps to improve recognition performance in re-identification and verification scenarios. To this end, we design a neural architecture for jointly embedding face and voice data, and we experiment with several training losses and audio-visual datasets. The idea is to exploit the relation between voice characteristics and facial morphology, so that face and voice uni-biometric models help each other to recognize people when trained jointly. Extensive experiments on four real-world datasets show that the biometric feature representation of a uni-biometric model jointly trained performs better than the one computed by the same uni-biometric model trained alone. Moreover, the recognition results are further improved by embedding face and voice data into a single shared representation of the two modalities. The proposed fusion strategy generalizes well on unseen and unheard users, and should be considered as a feasible solution that improves model performance. We expect that this chapter will support the biometric community to shape the research on deep audio-visual fusion in real-world contexts.

© Springer Nature Switzerland AG 2020
M. De Marsico et al. (Eds.): ICPRAM 2019, LNCS 11996, pp. 136–157, 2020.
https://doi.org/10.1007/978-3-030-40014-9_7

Keywords: Multi-biometric system · Cross-modal biometrics · Deep biometric fusion · Audio-visual learning · Verification · Re-identification

1   Introduction

Over the years, from visitors identified in human-robot interactions [27,28,46] to learners authenticated in online education platforms [13,14], biometrics has been increasingly playing a primary role in various contexts, such as robotics, medicine, science, engineering, education and several other business areas [25]. Evidence of this can be retrieved in recent reports that estimate a huge growth of the biometric market size, moving from \$10.74 billion in 2015 to \$32.73 billion by 2022[1]. Examples of biometric traits include the facial structure [49], the ridges of a fingerprint [35], the iris pattern [4], the sound waves of a voice [16], and the way a person interacts with a digital device [56]. From the system perspective, recognition pipelines detect the modality of interest in the biometric sample. This is followed by a set of pre-processing functions. Features are then extracted from pre-processed data, and used by a classifier for recognition. From the user perspective, an individual is asked to provide some samples whose feature vectors are stored as a template by the system (i.e., enrollment). Then, the recognition process may involve associating an identity with the probe (i.e., re-identification) or determining if the probe comes from the declared person (i.e., verification).

Most biometric systems manipulate a single modality (e.g., face only), and may encounter problems due to several factors surrounding the system and the user, such as the acquisition distance and the environmental conditions [3,10,44]. Deploying such systems in real-world scenarios thus presents various challenges. For instance, facial images exhibit large variations due to occlusions, pose, indoor-illumination, expressions, and accessories [49]. Similarly, audio samples vary due to the distance of the subject from the microphone, indoor reverberations, background noise, and so on [16]. Given the highly-variable conditions of these scenarios, focusing exclusively on one modality might seriously decrease the system reliability, especially when the acquisition conditions are not controlled. Multi-biometric systems have been proven to overcome some limitations of uni-biometric systems by combining evidence from different sources. This often results in improved recognition performance and enhanced system robustness, since the combined information is likely to be more distinctive compared to the one obtained from a single source [2,12]. Multi-biometric systems might be exploited in several scenarios, such as when people are speaking while being assisted by robots or when learners are attending an online oral exam.

One of the main design choices while developing a multi-biometric system is to select the level of the recognition pipeline where the fusion happens. To provide a unique global response, fusion policies generally refer to sensor level, feature level, score level, or decision level [41]. First, sensor-level fusion corresponds to combining raw data immediately after acquisition. Second, feature-level fusion refers to performing fusion of feature vectors extracted from different biometric

---

[1] https://www.grandviewresearch.com/industry-analysis/biometrics-industry.

samples. Third, score-level fusion corresponds to fuse matching scores produced by different systems. Fourth, decision-level fusion implies the combination of decisions taken by more than one system based on voting. Late fusion strategies usually made the process simpler and flexible, but an excessive amount of information entropy was lost. Early fusion policies were proven to work better, but tended to introduce high complexity and less flexibility. The recent revolution driven by deep-learned representations has contributed to reduce the latter deficiencies, and facilitated the experimentation of cost-effective intermediate fusion during model training and deployment [36]. It follows that, for instance, face and voice models might be trained jointly to learn whether face or voice probes come from a given user, but then deployed in a uni-biometric manner. On the other side, they could be combined in a multi-biometric way by embedding face and voice data into a single feature vector during deployment.

In this chapter, we introduce a multi-biometric training strategy that digests face and voice traits, and we investigate how it makes it possible to improve recognition performance in audio-visual re-identification and verification. With this in mind, we design a neural architecture composed by a sub-network for faces and a sub-network for voices fused at the top of the network and jointly trained. By exploiting features correlation, both models help each other to predict whether facial or vocal probes come from a given user. This paper extends the work presented in [30] that introduced an audio-visual dataset collected during human-robot interactions, and evaluated it and other challenging datasets on uni-biometric recognition tasks. The mission is to make a step forward towards the creation of biometric models able to work well on challenging real-world scenarios, such as identification performed by robots or continuous device authentication. More precisely, this paper provides the following contributions:

- We present a deeper contextualization of the state-of-the-art biometric solutions explored by researchers in audio-visual real-world biometrics scenarios.
- We experiment with a fusion strategy that combines face and voice traits instead of using them individually for re-identification and verification tasks.
- We extensively validate our strategy in public datasets, showing that it significantly improves uni-biometric and multi-biometric recognition accuracy.

Experiments on four datasets from real-world contexts show that the jointly-trained uni-biometric models reach significantly higher recognition accuracy than individually-trained uni-biometric models, both when their embeddings are deployed separately (i.e., 10%–20% of improvement) and when they are combined into a single multi-biometric embedding (i.e., 30%–50% of improvement). As the proposed strategy well generalizes on unseen and unheard users, it should be considered as a feasible solution for creating effective biometric models.

The rest of this chapter guides readers along the topic as follows. Section 2 summarizes recent uni-biometric strategies for face and voice recognition tasks together with biometric fusion strategies involving them. Then, Sect. 3 describes the proposed fusion strategy, including its formalization, input data formats, network architecture structures, and training process steps. Sections 4 depicts the experimental evaluation of the proposed strategy, and highlights how it outperforms state-of-the-art solutions. Finally, Sect. 5 depicts conclusions, open challenges and future directions in this research area.

2 Related Work

In this section, we briefly describe state-of-the-art contributions on audio-visual biometrics applied in different scenarios. This is achieved by introducing face and voice uni-biometric systems and, subsequently, existing multi-biometric models.

2.1 Deep Face Recognition

The recent widespread of deep learning in different areas has favoured the usage of neural networks as feature extractors combined with common machine-learning classifiers, as proposed in [50]. Backbone architectures that accomplish this task rapidly evolved from *AlexNet* [24] to *SENet* [20] over last years.

In parallel, researchers have formulated both data sampling strategies and loss functions to be applied when such backbone architectures are trained. *Deepface* [44] integrates a cross-entropy-based *Softmax* loss while training the network. However, applying *Softmax* loss is usually not sufficient by itself to learn features separated by a large margin when the samples come from diverse entities, and other loss functions have been explored to enhance the generalization ability. For instance, euclidean-distance-based losses embed images into an euclidean space and reduce intra-variance while enlarging inter-variance across samples. *Contrastive* loss [43] and *Triplet* loss [38] are commonly used to this end, but they often exhibit training instability and complex sampling strategies. *Center* loss [51] and *Ring* loss [55] balance the trade-off between accuracy and flexibility. Furthermore, cosine-margin-based losses, such as *AM-Softmax*, were proposed to learn features separable through angular distance measures [48].

2.2 Deep Voice Recognition

Traditional speaker recognition systems based on hand-crafted solutions relied on *Gaussian Mixture Models* (GMMs) [37] that are trained on low dimensional feature vectors, *Joint Factor Analysis* (JFA) [11] methods that model speaker and channel subspaces separately, or *i-Vectors* [23] that attempt to embed both subspaces into a single compact, low-dimensional space.

Modern systems leveraged deep-learned acoustic representations, i.e., embeddings, extracted from one of the last layers of a neural network trained for standard or one-shot speaker classification [19, 29]. The most prominent examples include *d-Vectors* [47], *c-Vectors* [7], *x-Vectors* [42], *VGGVox-Vectors* [34] and *ResNet-Vectors* [9]. Furthermore, deep learning frameworks with end-to-end loss functions to train speaker discriminative embeddings have recently drawn attention [18]. Their results proved that end-to-end systems with embeddings achieve better performance on short utterances common in several contexts (e.g., robotics, proctoring, and border controls) compared with hand-crafted systems.

2.3   Deep Audio-Visual Recognition

Combining signals from multiple sensors has been traditionally investigated from a data fusion perspective. For instance, such a merge step can happen at *sensor level* or *feature level*, and focuses on how to combine data from multiple sources, either by removing correlations between modalities or representing the fused data in a common subspace; the fused data is then fed into a machine-learning algorithm [1]. The literature provides also evidence of fusion techniques at *score level* and *decision level* [8,21,32,39]. There is no a general conclusion on which fusion policy performs better between early and late fusion, and the performance is problem-dependent [41]. However, late fusion was simpler to be implemented, particularly when modalities varied in dimensionality and sampling rates.

Emerging machine-learning strategies are making it possible to fill this gap in flexibility between early and late fusion. Through a new form of multi-biometric fusion of features representations, namely *intermediate fusion*, neural networks offer a flexible approach to multi-biometric fusion for numerous practical problems [36]. Given that neural architectures learn a hierarchical representation of the underlying data across its hidden layers, learned representations of different modalities can be fused at various levels of abstraction, introducing several advantages with respect to previous solutions [39,40]. Modality-wise and shared representations are learned from data, while features were originally manually designed and required prior knowledge on the data. Such a new fusion level requires little or no pre-processing of input data, differently from traditional techniques that may be sensitive to data pre-processing. Furthermore, implicit dimensionality reduction within the architecture and easily scalable capabilities are guaranteed, improving flexibility and accuracy at the same time.

Good evidence of these advantages comes from the literature. For instance, the authors in [15] aimed to learn features from audio and faces from convolutional neural networks compatible at high-level. Their strategy has been proven to produce better performance than single modality, showing the effectiveness of the multi-biometric fusion during deployment. The works in [5,6] proposed time-dependent audio-visual models adapted in an unsupervised fashion by exploiting the complementary of multiple modalities. Their approach allowed to control the model adaptation and to cope with situations when one of the two modalities is under-performing. Furthermore, the approach described in [45] used a three-dimensional convolutional neural network to map both modalities into a single representation space, and evaluated the correspondence of audio–visual streams using such learned multi-biometric features. Inspired by findings on high-level correlation of voice and face across humans, the authors in [40] experimented with an attention-based neural network that learns multi-sensory associations for user verification. The attention mechanism conditionally selects a salient modality representation between speech and facial ones, balancing between complementary inputs. Differently, the method in [52] extracted static and dynamic face and audio features; then, it concatenated the top discriminative visual-audio features to represent the two modalities, and used a linear classifier for identification. Recent experience in [26] depicted an efficient attention-guided audio-face fusion approach to detect speakers. Their factorized model deeply fused the paired audio-face features, whereby the joint audio-face representation can be reliably obtained. Finally, the authors in [33] investigated face-voice embeddings enabling cross-modal retrieval from voice to face and vice versa.

The collection of large amounts of training data and the advent of powerful graphics processing units (GPUs) is enabling deep intermediate fusion, and this paper makes a step forward towards its application in the audio-visual domain.

3   The Proposed Intermediate Fusion Approach

In this section, we describe our intermediate fusion strategy that jointly learns voice and face embeddings, including model formalization, input data formats, underlying architectures, and training details (Fig. 1).

The core idea is to leverage the morphological relations existing between voice and face biometrics in order to investigate a cross-modal training where each uni-biometric model is supported by the biometric model of the other modality in improving the effectiveness of its feature representations. Differently from other intermediate fusion approaches, such a multi-biometric fusion might happen (i) on training to develop better uni-biometric models and/or (ii) on deployment to exploit joint evidence from the two modalities simultaneously.

Face Backbone Formalization. Let $A_f \subset \mathbb{R}^{m \times n \times 3}$ denote the domain of RGB images with $m \times n \times 3$ size. Each image $a_f \in A_f$ is pre-processed in order to detect the bounding box and key points (two eyes, nose and two mouth corners) of the face. The affine transformation is used to align the face. The image is then resized and each pixel value is normalised in the range $[0, 1]$. The resulting intermediate facial image, defined as $S_f \subset \mathbb{R}^{m \times n \times 3}$, is used as input of the visual modality branch of our model. In this branch, an explicit feature extraction which produces fixed-length representations in $D_f \subset \mathbb{R}^e$. We denote such a stage as $\mathcal{D}_{f_{\theta_f}} : A_f \to D_f$. Its output is referred to as face feature vector.

Voice Backbone Formalization. Let $A_v \subset \mathbb{R}^*$ denote the domain of waveforms digitally represented by an intermediate visual acoustic representation $S_v \subset \mathbb{R}^{k \times *}$, such as a spectrogram or a filter-bank. Each audio $a_v \in A_v$ is converted to single-channel. The spectrogram is then generated in a sliding window fashion using a Hamming window, generating an acoustic representation s_v that corresponds to the audio a_v. Mean and variance normalisation is performed on every frequency bin of the spectrum. The resulting representation is used as input of the acoustic modality branch of our model. In this branch, an explicit feature extraction which produces fixed-length representations in $D_v \subset \mathbb{R}^e$. We denote such a stage as $\mathcal{D}_{v_{\theta_v}} : S_v \to D_v$. Its output is named voice feature vector.

Fusion Backbone Formalization. Let $D^{2 \times e}$ be the domain of audio-visual feature vectors generated by a plain concatenation of the sparse representation from the face and voice backbones, i.e., d_f and d_v. We denote as $\mathcal{C}_\theta : (D_f, D_v) \to D^{2 \times e}$ such a concatenation stage of both modalities applied after the representation layer of each single modality branch. Then, an additional feature vector learning

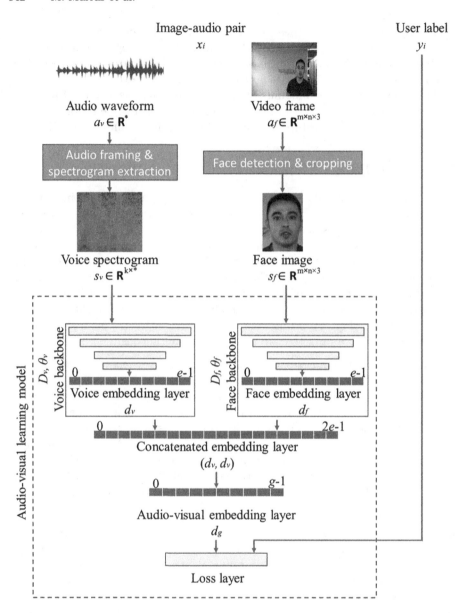

Fig. 1. The proposed neural architecture for intermediate multi-biometric fusion.

step is applied to the concatenated vector $d \in D^{2 \times e}$ to get a single feature vector of size g jointly learned from d_f and d_v. This extra layer aims to (i) keep independent the multi-biometric embedding size from the uni-biometric embedding sizes and (ii) learn more compacted and flexible representations. Moreover, by setting $g = e$, reasonable comparisons between uni-biometric and multi-biometric sparse representations of the same size can be performed. We denote such an extra step as $\mathcal{D}_{fv_{\theta_{f,v}}} : D^{2\times e} \to D^g$. Its output is named as audio-visual feature vector.

Combining both modalities might generate a better sparse representation of the individual, and enrich the feature representation of a single modality. This is due to the relations of voice to genre and facial morphology of people, e.g., male people commonly have a tone lower than female people. Therefore, by leveraging the fusion backbone, the uni-biometric backbones help each other to better recognize people. Our hypothesis is that the embeddings of each backbone should perform better when trained jointly than when trained separately.

Backbones Instantiation. The proposed approach makes use of existing neural network architectures, slightly arranged to accommodate the modality digested by each of the above-mentioned backbones and the subsequent fusion purposes.

Two instances of the residual-network (*ResNet-50*) architecture are used as feature vector extractors $\mathcal{D}_{f_{\theta_f}}$ and $\mathcal{D}_{v_{\theta_v}}$ within face and voice backbones, respectively [17]. Such a network, well known for good classification performance on visual and acoustic modalities [9,44], is similar to a multi-layer convolutional neural network, but with added skip connections such that the layers add residuals to an identity mapping on the channel outputs. The input layers of the original *ResNet-50* architecture are adapted to the modality associated to the corresponding each backbone. Moreover, the fully-connected layer at the top of the original network is replaced by two layers: a flatten layer and a fully-connected layer whose output is the embedding of the modality, i.e., d_f or d_v.

The fusion backbone $\mathcal{D}_{fv_{\theta_{f,v}}}$ is instantiated by a concatenation layer stacked into the model to combine face and voice feature vectors in $D^{2\times e}$ domain, and an additional fully-connected layer where the significant features of video and audio modality are jointly embedded. The latter output represents the audio-visual feature vector $d \in D^g$ previously formalized. Moreover, for each fully-connected layer, batch normalization has been set before the activation function to regularize the outputs, and a dropout layer is inserted after activation to prevent model over-fitting. Finally, an output layer depending on the applied loss function is posed at the top of the network during training.

Training Process Description. The training data is composed by N tuples $\{(x_i, y_i)\}_{i=1}^N$ where each multi-biometric sample x_i corresponds to a person associated with the class $y_i \in 1, ..., I$, being I the number of different identities depicted in N samples. Each sample x_i is defined as a pair $x_i = (a_{v_i}, a_{f_i})$ such that a_{v_i} is a utterance and a_{f_i} is a visual frame. The elements of each pair are randomly chosen among face and voice samples from the same user; then, they are sequentially fed into the multi-biometric model. Such a model can be integrated with any existing loss function. Additionally, a hold-out validation set consisting of all the speech and face segments from a single randomly-selected video per user is used to monitor training performance.

4   Experimental Evaluation

In this section, we assess the effectiveness of our fusion strategy. First, we detail the datasets, the experimental protocols, the implementation details, and the loss functions. Then, we present the results achieved by the fusion strategy on re-identification and verification, varying the loss function and the testing dataset.

4.1   Training and Testing Datasets

We considered traditional audio-visual datasets for training the models, and we tested them on datasets from diverse audio-visual contexts (see Fig. 2). This

a) VoxCeleb1

b) MOBIO

c) MSU-AVIS

d) AveRobot

Fig. 2. Facial samples coming from the testing datasets used to evaluate our approach.

choice enables the computation of additional state-of-the-art benchmark scores on *AveRobot*, and make it possible to observe how the strategy affects the performance on different contexts. The audio-visual datasets are divided in one training dataset and four testing datasets to replicate a cross-dataset setup:

- **Training Dataset.** *VoxCeleb1-Dev* is an audio-visual speaker identification and verification dataset collected by [34] from Youtube, including 21,819 videos from 1,211 identities. It is the one of the most suited for training a deep neural network due to the wide range of users and samples per user.
- **Testing Dataset #1.** *VoxCeleb1-Test* is an audio-visual speaker identification and verification dataset collected by [34] from Youtube, embracing 677 videos from 40 identities.
- **Testing Dataset #2.** *MOBIO* is a face and speaker recognition dataset collected by [31] from laptops and mobile phones under a controlled scenario, including 28,800 videos from 150 identities.
- **Testing Dataset #3.** *MSU-Avis* is a face and voice recognition dataset collected by [8] under semi-controlled indoor surveillance scenarios, including 2,260 videos from 50 identities.
- **Testing Dataset #4.** *AveRobot* is an audio-visual biometric recognition dataset collected under robot assistance scenarios in [30], including 2,664 videos from 111 identities.

The reader notices that acquisition distance, environmental conditions, and data quality greatly vary among the datasets, making them challenging.

4.2 Evaluation Setup and Protocols

Experiments aimed to assess both uni-biometric and multi-biometric feature representations through evaluation protocols applied in re-identification and verification tasks (Fig. 3).

Tested Data Format. For the face branch, each frame is analyzed in order to detect the face area and landmarks through MTCNN [53]. The five facial points (two eyes, nose and two mouth corners) are adopted by such an algorithm to perform face alignment. The faces are then resized to 112×112 pixels in order to fit in our branch and each pixel in *[0, 255]* in RGB images is normalized by subtracting *127.5* then dividing by *128*. The resulting images are then used as input to the face branch. For the voice branch, each audio is converted to single-channel, *16*-bit streams at a *16* kHz sampling rate for consistency. The spectrograms are then generated in a sliding window fashion using a Hamming window of width 25 ms and step 10 ms. This gives spectrograms of size 512×300 for three seconds of speech. Mean and variance normalisation is performed on every frequency bin of the spectrum. No other speech-specific pre-processing is used. The spectrograms are used as input to the voice branch.

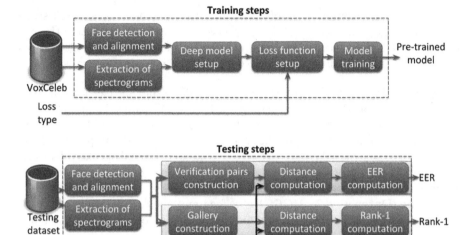

Fig. 3. Experimental evaluation overview. Training and testing protocols.

Tested Feature Representations. The evaluation involved uni-biometric and multi-biometric feature representations obtained from backbone networks trained on top of *VoxCeleb1-Dev*. In order to optimize model weights, several instances of the network were independently trained through different loss functions from various families: *Softmax* loss [44], *Center* loss [51], *Ring* loss [55], and *AM-Softmax* loss [48]. More precisely, for each training loss, we trained appropriate models to learn the following feature representations:

- *Uni-Modal Voice* representations extracted from d_v when the voice branch is trained alone (baseline).
- *Uni-Modal Face* representations extracted from d_f when the face branch is trained alone (baseline).
- *Multi-Modal Voice* representations extracted from d_v when the voice branch is trained jointly with the face branch (introduced in this paper).
- *Multi-Modal Face* representations extracted from d_f when the face branch is trained jointly with the voice branch (introduced in this paper).
- *Multi-Modal Face+Voice* representations extracted from d_g when the face branch and the voice branch are jointly trained (introduced in this paper).

Each model was initialised with weights pre-trained on ImageNet. Stochastic gradient descent with a weight decay set to *0.0005* was used on mini-batches of size *512* along *40* epochs. The initial learning rate was *0.1*, and this was decreased with a factor of *10* after *20*, *30* and *35* epochs. The training procedure was coded in Python, using Keras on top of Tensorflow.

Re-identification Protocol. For each testing dataset, the protocol aims to evaluate how the learned representation are capable of predicting, for a given test frame/spectrogram, the identity of the person chosen from a gallery of identities. For each experiment conducted on a testing dataset, we randomly selected 40 users every time in order to (i) keep constant the number of considered users, and (ii) maintain comparable the results across the different datasets. *VoxCeleb1-Test* has the minimum number of participants among the considered datasets (i.e., *40*). For each user, we have chosen the first *80%* of videos for the gallery, while the other *20%* of videos were probes. For each user, we randomly selected *20* frames/spectrograms from the gallery videos as gallery images, and *100* frames/spectrograms from the probe videos as probe images. Then, given each frame/spectrogram, the corresponding feature representation was extracted. The *Euclidean* distance was used to compare feature vectors obtained from models trained on *Softmax*, *Center* loss and *Ring* loss, while the *Cosine* distance was used for features vectors obtained from models trained on *AM-Softmax* loss due to its underlying design. Then, we measured the top one rank, a well-accepted measure to evaluate the performance on people re-identification tasks (e.g., [54]). The probe image is matched against a set of gallery images, obtaining a ranked list according to their matching similarity/distance. The correct match is assigned to one of the top ranks, the top one rank in this case (*Rank-1*).

The *Rank-1* is formulated as the accuracy on predicting the right identity (prediction) given the known spectrogram/face identity (ground truth):

$$Rank\text{-}1 = \frac{TP + TN}{TP + TN + FP + FN} \tag{1}$$

where TP is the true positive, TN represents the true negative, FP is the false positive and FN represents the false negatives. Thus, it was used to evaluate the performance of the models on the test images/spectrograms. Starting from the subject selection, the experiment was repeated and the results were averaged.

Verification Protocol. For each testing dataset, the protocol aims to evaluate how the learned representations are capable of verifying, given a pair of test frames/spectrograms, whether the faces/voices come from the same person. From each testing dataset, we randomly selected *40* subjects due to the same reasons stated in the above re-identification protocol. Then, we randomly created a list of *20* videos (with repetitions) for each selected user and, from each one of them, we randomly created *20* positive frame pairs and *20* negative frame pairs. The above-mentioned feature representations were considered as feature vector associated to each frame/spectrogram. We used the same distance measures leveraged for re-identification and the *Equal Error Rate* (EER) was computed to evaluate the performance of the models on the test pairs. *EER* is a well-known biometric security metric measured on verification tasks [22]. *EER* indicates that the proportion of false acceptances (*FAR*) is equal to the proportion of false rejections (*FRR*). Both measures are formulated as:

$$FAR = \frac{\text{number of false accepts}}{\text{number of impostors comparisons}}$$
$$FRR = \frac{\text{number of false rejects}}{\text{number of genuine comparisons}} \tag{2}$$

The lower the EER, the higher the performance. Lastly, starting from the subject selection, the experiment was repeated and the results were averaged.

4.3 Re-Identification Results

The *Rank-1* performance on the testing datasets is shown in Figs. 4, 5, 6 and 7. It can be observed that the results vary with respect to the modality, training loss, and the dataset. Results are presented from these three point of views.

Considering the face modality, the representations learned through the *Softmax* loss appear as the best performer for the uni-modal face setup (*Rank-1* from *0.32* to *0.74*), while the representations performance for the multi-modal face setup greatly varies among the training losses and the testing datasets. This means that the deep multi-biometric training strategy is strongly affected by the training loss and the targeting dataset, while common uni-biometric training strategies take advantage of the *Softmax* loss and their results are affected only by the dataset. Furthermore, it can be observed that multi-modal face representations make it possible to improve the results in face re-identification on challenging scenarios as in the *AveRobot* dataset. In more controlled scenarios, while the deep fusion allows us to increase the accuracy, it reaches results comparable with the ones obtained by uni-modal features representations learned through *Softmax* loss in uni-biometric face models.

Different observations can be made for the voice modality. The representations learned through the *Center* loss are superior to representations learned by other losses in uni-modal and multi-modal voice models. Interestingly, the multi-modal voice representations perform worse than the uni-modal voice representations for any loss. It follows that voice biometrics does not take a large advantage of the deep fusion strategy, differently from what happens for face biometrics. The exception is represented by results obtained in *MOBIO* multimodal voice representations; they reach higher results than the uni-modal voice representations. This means that there is not a general conclusion regarding the effectiveness of multi-modal voice representations, but they are datasetdependent. Therefore, preliminary tests should be performed to select the right voice training strategy based on the context.

In the case both face and voice biometrics are fused (*multi-modal face+voice* setup), the representations learned through *Ring* and *Center* losses achieved better results than uni-modal representations, while the representations learned through *Softmax* and *AM-Softmax* losses reach worse results probably due to the bad performance of the intermediate multi-modal voice representation. It follows that a plain fusion of face and voice embeddings during training is not always sufficient to improve the results with respect to uni-modal representations. It appears necessary to design countermeasures for controlling the contribution of each modality on the audio-visual embedding during training.

Among the datasets, *VoxCeleb1-Test* showed the highest *Rank-1* values. This is probably related to the fact that all the models are trained on data coming from the same context of the testing dataset. On *MOBIO*, the representations tended to achieve comparable results as the data includes front-face videos recorded

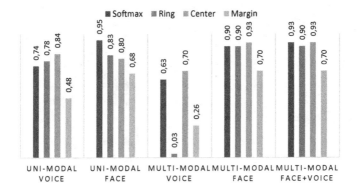

Fig. 4. Re-identification results on VoxCeleb1-Test - Rank-1.

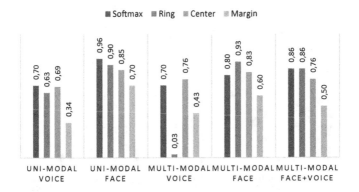

Fig. 5. Re-identification results on MOBIO - Rank-1.

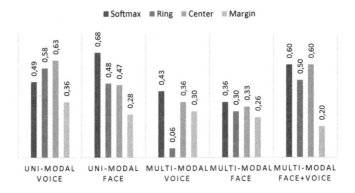

Fig. 6. Re-identification results on MSU-Avis - Rank-1.

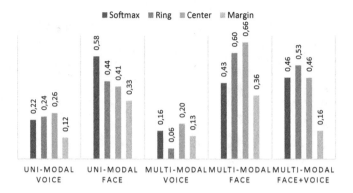

Fig. 7. Re-identification results on Averobot - Rank-1.

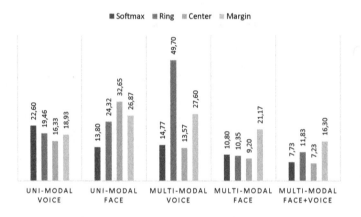

Fig. 8. Verification results on VoxCeleb1-Test - EER.

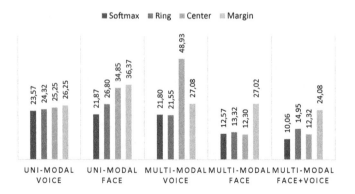

Fig. 9. Verification results on MOBIO - EER.

from the smartphone, i.e., a controlled conditions where the recognition should be easier. Differently, *MSU-Avis* and *AveRobot* highlight several challenges for the trained representations. The more uncontrolled scenarios conveyed by the latter datasets are the main reasons of the significantly lower *Rank-1* values. In particular, the *AveRobot* dataset represents the most challenging scenario, and more effective fusion strategies should be designed starting from the one presented in this paper.

4.4   Verification Results

Figures 8, 9, 10 and 11 plot the results achieved by the learned representations on verification. The ranking is slightly different with respect to the re-identification task, and the impact of the context, the loss, and the modality varies across settings.

It can be observed that multi-modal face representations achieve lower *EER* than uni-modal face representations with all the dataset and training losses. This means that deep fusion significantly helps to create better sparse representations for the face modality. More precisely, *EER* obtained by representations learned through *Ring* and *Center* losses can be improved of around *50%*, while we observed an improvement of around *25%* thanks to representations learned through *Softmax* and *Margin* losses. It follows that multi-modal face sparse representations better separate among genuine and impostor pairs.

Comparable results are obtained by multi-modal voice representations, even though the improvement with respect to the uni-modal voice representations is less evident, i.e. among *5%* and *10%*. Interestingly, multi-modal voice representations learned through *Ring* loss do not work well. It follows that, as shown on the re-identification task, the *Ring* loss suffers from the deep fusion approach. Our results suggest that such a loss has a minor impact in deep audio-visual fusion settings.

By merging face and voice embeddings into a single representation, the verification performance improves on all the datasets, with all the training losses. It can be observed an improvement of around *50%* on all the settings. The face-voice fused representations work well also when learned through *Ring* loss; hence, the deficiencies experienced by multi-modal voice representations learned through *Ring* loss are mitigated by fusing voices and faces.

The results across the testing datasets confirm the observations made for the re-identification task. The context has a relevant impact on the absolute performance of the models, moving from *VoxCeleb1-Test* to *AveRobot* by increasing challenging level. In particular, the verification results on *AveRobot* pairs are *4* or *5* times worse than the ones achieved on *VoxCeleb1-Test* pairs. The reasons behind this large difference could be related to the more uncontrolled conditions characterized by very dark faces and highly-noisy surroundings.

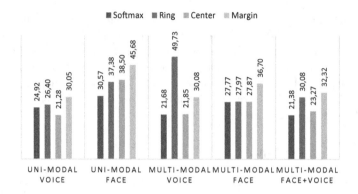

Fig. 10. Verification results on MSU-Avis - EER.

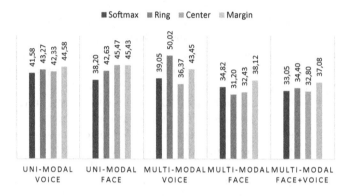

Fig. 11. Verification results on Averobot - EER.

5   Conclusions, Open Challenges, and Future Directions

In this chapter, we proposed a deep intermediate fusion strategy of audio-visual biometric data. By combining state-of-the-art deep learning methodologies, a two-branch neural network fed with face and voice pairs aimed to jointly learn uni-biometric and multi-biometric fixed-length feature representations by exploiting feature correlation. Branches influence each other in computing the right classification label after their fusion during training, so that the representation layer of each uni-biometric model performs better than the one returned by a uni-biometric model trained alone. The results were further improved by jointly learning a single audio-visual embedding that includes information from both face and voice evidence. Based on the obtained results, we can conclude that:

– Face and voice models can benefit from deep intermediate fusion, and the recognition improvement depends on the modality, the loss, and the context.
– Deep intermediate fusion during training can be used to significantly increase recognition accuracy of uni-biometric face and voice models.

- Uni-biometric face models exhibit higher accuracy improvements than uni-biometric voice models after being jointly trained at intermediate level.
- Merging face and voice into a single embedding vector at intermediate level positively impacts the accuracy of multi-biometric audio-visual models.
- Face and voice models jointly trained at intermediate level generalize well across populations and are more robust when applied in challenging contexts.
- Deep intermediate fusion should be considered as a viable solution for creating more robust and reliable biometric models.

Research on deep multi-biometric fusion has produced a variety of solid methods, but still poses some interesting challenges that require further investigation:

- **Deep Fusion Capability.** Techniques in deep multi-modal learning facilitate a flexible intermediate-fusion approach, which not only makes it simpler to fuse modality-wise representations and learn a joint representation but also allows multi-modal fusion at various depths in the architecture. Moreover, deep learning architectures still involve a great deal of manual design, and experimenters may not have explored the full space of possible fusion architectures. It is natural that researchers should extend the notion of learning to architectures adaptable to a specific task.
- **Transferability across Contexts.** Existing models tend to be sensitive to the context targeted by the underlying training data. This has favored the creation of biometric models that, after being trained with data from a given context, do not generalize well in other contexts. With the new availability of public datasets and pre-trained models, it will become easier to plug them into a task different from the original one. Researchers could fine-tune pre-trained models with small amounts of context-specific data.
- **Robustness in Uncontrolled Environments.** Devising audio-visual biometric systems that can operate in unconstrained sensing environments is another unsolved problem. Most biometric systems either implicitly or explicitly impose some constraints on the data acquisition. Such constraints have to be reduced in order to seamlessly recognize individuals, i.e., the interaction between an individual and a biometric system should be transparent. This necessitates innovative interfaces and robust data processing algorithms.
- **Robustness against Spoofing Attacks.** Synthetically generated traits or maliciously modified traits are used to circumvent biometric systems. The challenge is to develop counter-measures that are applicable to hither to unseen or unknown attacks. Evaluating and assessing how the deployment of multi-biometric systems might help to face this challenge requires further investigation.
- **Explainablity and Interpretability.** Most machine-learning algorithms built into automation and artificial intelligence systems lack transparency, and may contain an imprint of the unconscious biases of the data and algorithms underlying them. Hence, it becomes important to understand how we can predict what is going to be predicted, given a change in input or algorithmic parameters. Moreover, it requires attention how the internal mechanics of the system can be explained in human terms.

- **Fairness, Transparency and Accountability.** With the advent of machine-learning, addressing bias within biometric systems will be a core priority due to several reasons. For instance, some biases can be introduced by using training data which is not an accurate sample of the target population or is influenced by socio-cultural stereotypes. Moreover, the methods used to collect or measure data and the algorithms leveraged for predicting identities can propagate biases. Future research should control these biases in the developed models, promoting fair, transparent, and accountable systems.

We expect that the case study on audio-visual fusion covered in this chapter will help researchers and developers to shape future research in the field.

Acknowledgments. Mirko Marras gratefully acknowledges Sardinia Regional Government for the financial support of his PhD scholarship (P.O.R. Sardegna F.S.E. Operational Programme of the Autonomous Region of Sardinia, European Social Fund 2014–2020, Axis III "Education and Training", Thematic Goal 10, Priority of Investment 10ii, Specific Goal 10.5).

This research work has been partially supported by the Spanish Ministry of Economy and Competitiveness (TIN2015-64395-R MINECO/FEDER) and the Spanish Ministry of Science, Innovation and Universities (RTI2018-093337-B-I00), by the Office of Economy, Industry, Commerce and Knowledge of the Canary Islands Government (CEI2018-4), and the Computer Science Department at the Universidad de Las Palmas de Gran Canaria.

References

1. Abozaid, A., Haggag, A., Kasban, H., Eltokhy, M.: Multimodal biometric scheme for human authentication technique based on voice and face recognition fusion. Multimed. Tools Appl. **78**, 1–17 (2018)
2. Barra, S., Casanova, A., Fraschini, M., Nappi, M.: Fusion of physiological measures for multimodal biometric systems. Multimed. Tools Appl. **76**(4), 4835–4847 (2017)
3. Barra, S., De Marsico, M., Galdi, C., Riccio, D., Wechsler, H.: FAME: face authentication for mobile encounter. In: 2013 IEEE Workshop on Biometric Measurements and Systems for Security and Medical Applications (BIOMS), pp. 1–7. IEEE (2013)
4. Bowyer, K.W., Burge, M.J.: Handbook of Iris Recognition. Springer, Heidelberg (2016). https://doi.org/10.1007/978-1-4471-6784-6
5. Brutti, A., Cavallaro, A.: Online cross-modal adaptation for audio-visual person identification with wearable cameras. IEEE Trans. Hum.-Mach. Syst. **47**(1), 40–51 (2016)
6. Cavallaro, A., Brutti, A.: Audio-visual learning for body-worn cameras. In: Multimodal Behavior Analysis in the Wild, pp. 103–119. Elsevier (2019)
7. Chen, Y.H., Lopez-Moreno, I., Sainath, T.N., Visontai, M., Alvarez, R., Parada, C.: Locally-connected and convolutional neural networks for small footprint speaker recognition. In: Sixteenth Annual Conference of the International Speech Communication Association (2015)
8. Chowdhury, A., Atoum, Y., Tran, L., Liu, X., Ross, A.: MSU-AVIS dataset: fusing face and voice modalities for biometric recognition in indoor surveillance videos. In: 2018 24th International Conference on Pattern Recognition (ICPR), pp. 3567–3573. IEEE (2018)

9. Chung, J.S., Nagrani, A., Zisserman, A.: VoxCeleb2: deep speaker recognition. arXiv preprint arXiv:1806.05622 (2018)
10. Cruz, C., Sucar, L.E., Morales, E.F.: Real-time face recognition for human-robot interaction. In: 8th IEEE International Conference on Automatic Face and Gesture Recognition, FG 2008, pp. 1–6. IEEE (2008)
11. Dehak, N., Kenny, P.J., Dehak, R., Dumouchel, P., Ouellet, P.: Front-end factor analysis for speaker verification. IEEE Trans. Audio Speech Lang. Process. **19**(4), 788–798 (2011)
12. Fenu, G., Marras, M.: Leveraging continuous multi-modal authentication for access control in mobile cloud environments. In: Battiato, S., Farinella, G.M., Leo, M., Gallo, G. (eds.) ICIAP 2017. LNCS, vol. 10590, pp. 331–342. Springer, Cham (2017). https://doi.org/10.1007/978-3-319-70742-6_31
13. Fenu, G., Marras, M.: Controlling user access to cloud-connected mobile applications by means of biometrics. IEEE Cloud Comput. **5**(4), 47–57 (2018)
14. Fenu, G., Marras, M., Boratto, L.: A multi-biometric system for continuous student authentication in e-learning platforms. Pattern Recogn. Lett. **113**, 83–92 (2018)
15. Geng, J., Liu, X., Cheung, Y.M.: Audio-visual speaker recognition via multi-modal correlated neural networks. In: 2016 IEEE/WIC/ACM International Conference on Web Intelligence Workshops (WIW), pp. 123–128. IEEE (2016)
16. Hansen, J.H., Hasan, T.: Speaker recognition by machines and humans: a tutorial review. IEEE Signal Process. Mag. **32**(6), 74–99 (2015)
17. He, K., Zhang, X., Ren, S., Sun, J.: Deep residual learning for image recognition. In: Proceedings of the IEEE Conference on Computer Vision and Pattern Recognition, pp. 770–778 (2016)
18. Heigold, G., Moreno, I., Bengio, S., Shazeer, N.: End-to-end text-dependent speaker verification. In: 2016 IEEE International Conference on Acoustics, Speech and Signal Processing (ICASSP), pp. 5115–5119. IEEE (2016)
19. Hershey, S., et al.: CNN architectures for large-scale audio classification. In: 2017 IEEE International Conference on Acoustics, Speech and Signal Processing (ICASSP), pp. 131–135. IEEE (2017)
20. Hu, J., Shen, L., Sun, G.: Squeeze-and-excitation networks. arXiv preprint arXiv:1709.01507 7 (2017)
21. Huang, L., Yu, C., Cao, X.: Bimodal biometric person recognition by score fusion. In: 2018 5th International Conference on Information Science and Control Engineering (ICISCE), pp. 1093–1097. IEEE (2018)
22. Jain, A., Hong, L., Pankanti, S.: Biometric identification. Commun. ACM **43**(2), 90–98 (2000)
23. Kanagasundaram, A., Vogt, R., Dean, D.B., Sridharan, S., Mason, M.W.: I-vector based speaker recognition on short utterances. In: Proceedings of the 12th Annual Conference of the International Speech Communication Association, pp. 2341–2344. International Speech Communication Association (ISCA) (2011)
24. Krizhevsky, A., Sutskever, I., Hinton, G.E.: ImageNet classification with deep convolutional neural networks. In: Advances in Neural Information Processing Systems, pp. 1097–1105 (2012)
25. Li, S.Z., Jain, A.: Encyclopedia of Biometrics. Springer, Heidelberg (2015)
26. Liu, X., Geng, J., Ling, H., Cheung, Y.M.: Attention guided deep audio-face fusion for efficient speaker naming. Pattern Recogn. **88**, 557–568 (2019)
27. López, J., Pérez, D., Santos, M., Cacho, M.: GuideBot. A tour guide system based on mobile robots. Int. J. Adv. Robot. Syst. **10**, 381 (2013)
28. López, J., Pérez, D., Zalama, E., Gomez-Garcia-Bermejo, J.: BellBot - a hotel assistant system using mobile robots. Int. J. Adv. Robot. Syst. **10**, 40 (2013)

29. Lukic, Y., Vogt, C., Dürr, O., Stadelmann, T.: Speaker identification and clustering using convolutional neural networks. In: 2016 IEEE 26th International Workshop on Machine Learning for Signal Processing (MLSP), 13–16 September 2016, Vietri sul Mare, Italy. IEEE (2016)

30. Marras, M., Marín-Reyes, P.A., Lorenzo-Navarro, J., Castrillón-Santana, M., Fenu, G.: AveRobot: an audio-visual dataset for people re-identification and verification in human-robot interaction. In: International Conference on Pattern Recognition Applications and Methods (2019)

31. McCool, C., et al.: Bi-modal person recognition on a mobile phone: using mobile phone data. In: 2012 IEEE International Conference on Multimedia and Expo Workshops (ICMEW), pp. 635–640. IEEE (2012)

32. Memon, Q., AlKassim, Z., AlHassan, E., Omer, M., Alsiddig, M.: Audio-visual biometric authentication for secured access into personal devices. In: Proceedings of the 6th International Conference on Bioinformatics and Biomedical Science, pp. 85–89. ACM (2017)

33. Nagrani, A., Albanie, S., Zisserman, A.: Learnable PINs: cross-modal embeddings for person identity. In: Proceedings of the European Conference on Computer Vision (ECCV), pp. 71–88 (2018)

34. Nagrani, A., Chung, J.S., Zisserman, A.: VoxCeleb: a large-scale speaker identification dataset. arXiv preprint arXiv:1706.08612 (2017)

35. Peralta, D., et al.: A survey on fingerprint minutiae-based local matching for verification and identification: taxonomy and experimental evaluation. Inf. Sci. **315**, 67–87 (2015)

36. Ramachandram, D., Taylor, G.W.: Deep multimodal learning: a survey on recent advances and trends. IEEE Signal Process. Mag. **34**(6), 96–108 (2017)

37. Reynolds, D.A., Quatieri, T.F., Dunn, R.B.: Speaker verification using adapted gaussian mixture models. Digit. Signal Proc. **10**(1–3), 19–41 (2000)

38. Schroff, F., Kalenichenko, D., Philbin, J.: FaceNet: a unified embedding for face recognition and clustering. In: Proceedings of the IEEE Conference on Computer Vision and Pattern Recognition, pp. 815–823 (2015)

39. Sell, G., Duh, K., Snyder, D., Etter, D., Garcia-Romero, D.: Audio-visual person recognition in multimedia data from the IARPA Janus program. In: 2018 IEEE International Conference on Acoustics, Speech and Signal Processing (ICASSP), pp. 3031–3035. IEEE (2018)

40. Shon, S., Oh, T.H., Glass, J.: Noise-tolerant audio-visual online person verification using an attention-based neural network fusion. In: ICASSP 2019–2019 IEEE International Conference on Acoustics, Speech and Signal Processing (ICASSP), pp. 3995–3999. IEEE (2019)

41. Singh, M., Singh, R., Ross, A.: A comprehensive overview of biometric fusion. Inf. Fusion **52**, 187–205 (2019)

42. Snyder, D., Garcia-Romero, D., Sell, G., Povey, D., Khudanpur, S.: X-vectors: robust DNN embeddings for speaker recognition. In: 2018 IEEE International Conference on Acoustics, Speech and Signal Processing (ICASSP), pp. 5329–5333. IEEE (2018)

43. Sun, Y., Liang, D., Wang, X., Tang, X.: DeepID3: face recognition with very deep neural networks. arXiv preprint arXiv:1502.00873 (2015)

44. Taigman, Y., Yang, M., Ranzato, M., Wolf, L.: DeepFace: closing the gap to human-level performance in face verification. In: Proceedings of the IEEE Conference on Computer Vision and Pattern Recognition, pp. 1701–1708 (2014)

45. Torfi, A., Iranmanesh, S.M., Nasrabadi, N., Dawson, J.: 3D convolutional neural networks for cross audio-visual matching recognition. IEEE Access **5**, 22081–22091 (2017)
46. Troniak, D., et al.: Charlie rides the elevator-integrating vision, navigation and manipulation towards multi-floor robot locomotion. In: 2013 International Conference on Computer and Robot Vision (CRV), pp. 1–8. IEEE (2013)
47. Variani, E., Lei, X., McDermott, E., Moreno, I.L., Gonzalez-Dominguez, J.: Deep neural networks for small footprint text-dependent speaker verification. In: 2014 IEEE International Conference on Acoustics, Speech and Signal Processing (ICASSP), pp. 4052–4056. IEEE (2014)
48. Wang, F., Cheng, J., Liu, W., Liu, H.: Additive margin softmax for face verification. IEEE Signal Process. Lett. **25**(7), 926–930 (2018)
49. Wang, M., Deng, W.: Deep face recognition: a survey. arXiv preprint arXiv:1804.06655 (2018)
50. Wang, Y., Shen, J., Petridis, S., Pantic, M.: A real-time and unsupervised face re-identification system for human-robot interaction. Pattern Recogn. Lett. **128**, 559–568 (2018)
51. Wen, Y., Zhang, K., Li, Z., Qiao, Y.: A discriminative feature learning approach for deep face recognition. In: Leibe, B., Matas, J., Sebe, N., Welling, M. (eds.) ECCV 2016. LNCS, vol. 9911, pp. 499–515. Springer, Cham (2016). https://doi.org/10.1007/978-3-319-46478-7_31
52. Zhang, J., Richmond, K., Fisher, R.B.: Dual-modality talking-metrics: 3D visual-audio integrated behaviometric cues from speakers. In: 2018 24th International Conference on Pattern Recognition (ICPR), pp. 3144–3149. IEEE (2018)
53. Zhang, K., Zhang, Z., Li, Z., Qiao, Y.: Joint face detection and alignment using multitask cascaded convolutional networks. IEEE Signal Process. Lett. **23**(10), 1499–1503 (2016). https://doi.org/10.1109/LSP.2016.2603342
54. Zheng, W.S., Gong, S., Xiang, T.: Reidentification by relative distance comparison. IEEE Trans. Pattern Anal. Mach. Intell. **35**(3), 653–668 (2013)
55. Zheng, Y., Pal, D.K., Savvides, M.: Ring loss: convex feature normalization for face recognition. In: Proceedings of the IEEE Conference on Computer Vision and Pattern Recognition, pp. 5089–5097 (2018)
56. Zhong, Y., Deng, Y.: A survey on keystroke dynamics biometrics: approaches, advances, and evaluations. In: Recent Advances in User Authentication Using Keystroke Dynamics Biometrics, pp. 1–22 (2015)

Author Index

Printed in the United States
By Bookmasters